1/10/17

NEW HORIZONS

REDISCOVERING PLUTO

**Marcia Dunn and
The Associated Press**

Mango Media
Miami
in collaboration with
The Associated Press

 AP EDITIONS

AP Editions

Published by Mango Media, Inc.
www.mangomedia.us

This is a work of non-fiction adapted from articles and content by journalists of The Associated Press and published with permission.

NEW HORIZONS *Rediscovering Pluto*

ISBN: 978-1-63353-243-4

Publisher's Note

AP Editions brings together stories and photographs by the professional journalists of The Associated Press.

These stories are presented in their original form and are intended to provide a snapshot of history as the moments occurred.

We hope you enjoy these selections from the front lines of news-gathering.

"I'm a little biased, but I think the solar system saved the best for last."

—Alan Stern, Southwest Research Institute planetary scientist.

Table of Contents

PREFACE

Who knew Pluto had a heart?

We're talking the tiny icy orb out in the solar system boon-docks, not the Disney pup that, coincidentally or not, made its debut the same year.

Now, 85 years after its discovery, Pluto the dwarf planet has been uncloaked. The views — peachy heart-shaped region and all — are astounding.

Rocky Mountain-size peaks sculpted of ice. Vast frozen plains featuring smooth hills and fields of small pits. Glacier-like ice flows spilling into impact craters and forming ponds of frozen nitrogen. Fat layers of haze stretching nearly 100 miles into the nitrogen-rich atmosphere. Tantalizing hints of icy volcanoes, geysers and, possibly, an underground ocean.

Yes, Pluto might still be geologically active.

NASA's New Horizons spacecraft provided Earthlings' first up-close look at Pluto in the summer of 2015, buzzing the smaller-than-our-moon world in a 31,000-mph flyby. The action unfolded 3 billion miles away following an epic journey of 9½ years, the space agency's last planetary stop on a solar-system tour that began more than a half-century ago.

No bigger than a baby grand piano, New Horizons was the first robotic explorer to ever visit Pluto. The spacecraft passed within 7,700 miles of Pluto on July 14, 2015.

Scientists were so intent on making the most of this once-in-a-generation-or-more event that they ceased communication with the craft during the flyby to maximize the observations. (The craft was designed to either talk with Earth or peer at Pluto, but not at the same time.) For 13 hours, the New Horizons team held its collective breath, wondering if the spacecraft had, No. 1, survived and, No. 2, gotten what it came for — a slew of unprec-edented pictures and measurements.

Team members were rewarded with an emphatic "Yes!" on both counts.

The planetary scientist leading the effort, Alan Stern, was so jazzed by the early results that he cautioned at a news confer-ence, "If you're seeing a cardiologist, you may want to leave the room. There are some pretty mind-blowing discoveries."

"Just feast your eyes on this," Stern said, displaying a haunt-ingly beautiful picture of a gold-and copper-hued Pluto alongside grayish jumbo moon Charon. "We've never been to a double-planet system before, and it's turning out to just really be a scientific wonderland."

The bright heart-shaped feature now bears the last name of Pluto's 1930 discoverer, American astronomer Clyde Tombaugh. A smidgen of his cremated remains, by the way, is aboard the spacecraft.

Tombaugh Regio, as the heart is affectionately known, is 1,000 miles across at its widest point and located just above Pluto's equator, in the Northern Hemisphere. That's where the well-documented plains are located and, just on the border, the mountain ranges.

The heart is massive, considering that Pluto's diameter spans 1,473 miles, 12 miles give or take, or about two-thirds the size of Earth's moon. (We can thank New Horizons for that up-dated diameter measurement, bigger than previous estimates.)

The two "lobes" of the heart appear to be distinctly different, and scientists theorize nitrogen snow could be blowing from the brighter west side to the east.

Not to be outdone, Charon, about half Pluto's size, also was scrutinized during the encounter, as were Pluto's four baby-size moons — Nix, Hydra, Kerberos and Styx. (All names pertain to the mythological underworld.)

New Horizons passed within 18,000 miles of Charon, re-vealing incredibly deep canyons.

The spacecraft returned fuzzy pictures of potato-shaped Nix and Hydra, barely big enough to accommodate a 26.2-mile mar-athon. The team is still awaiting decent snapshots of even smaller Kerberos and Styx.

Somewhat to Stern's surprise, no further moons have mate-rialized. The Hubble Space Telescope uncovered all four of Pluto's mini-moons over the past decade, and if any other moons were out there, New Horizons likely would have spotted them by now.

Until New Horizons, Pluto and Charon were merely blurry blobs from afar. Hubble offered the best views, and they weren't much to brag about. In the coming months, many more revealing pictures are expected as New Horizons unloads its precious stash of information.

New Horizons collected so much data that it will take until fall of 2016 to relay everything to flight controllers. Signals take 4½ hours, one way, to travel between New Horizons and Mission Control at Johns Hopkins University's Applied Physics Laboratory in Laurel, Maryland.

By mid-August, one month after the historic encounter, New Horizons already was 25 million miles beyond Pluto and heading ever farther afield.

Managers are rooting for a mission extension so New Horizons can swing by another iceball in this so-called Twilight Zone, or Third Zone, region of the solar system. Officially, it's known as the Kuiper Belt, home to comets and other frosty objects galore.

The managers also are hoping for a redo of Pluto's planetary status.

Pluto was considered a full-fledged planet when New Horizons rocketed away from Cape Canaveral, Florida, on January 19, 2006. At the time, it was No. 9 in the solar-system lineup and the last planet to be sent a visitor. Just seven months into the flight, however, Pluto was demoted to dwarf planethood on various technicalities.

As the media hubbub swirled before, during and immediately after the July flyby — New Horizons was a social media sensation — Stern and others repeatedly flashed the Pluto salute. Nine fingers up. Nine for planet No. 9.

The New Horizons team also wasted no time retooling the Pluto bumper stickers. They now show Pluto, heart prominent, in a car's rearview mirror. The bumper sticker reads: "My other vehicle explored Pluto."

Explored, past tense.

Unexplored no more.

By Marcia Dunn

Aerospace Writer Marcia Dunn with the space shuttle Atlantis in the background at Kennedy Space Center in Cape Canaveral, Florida. She has covered the space program for the AP since 1990. (AP Photo/John Raoux)

Chapter 1

CLYDE TOMBAUGH

Pluto at 85: From Discovery to New Horizons at Lowell Observatory in Flagstaff, Arizona, a yearlong exhibit celebrates the work of amateur astronomer Clyde Tombaugh, who discovered Pluto in 1930, March 13, 2015. (AP Photo/Felicia Fonseca)

PLANET X
Chicago, Tuesday, January 22, 1980

Fifty years ago, a 24-year-old former wheat farmer peered far out into the solar system and electrified the science world by discovering a new planet.

Clyde Tombaugh named his discovery Pluto, after the god of the underworld.

"It was a big break for me," he recalled. "At the time I had no college education, only high school."

Tombaugh had studied the stars since he was 12, growing up in rural Illinois and later Kansas. When times got tough on the farm, he took a job at the Lowell Observatory in Flagstaff, Arizona. It was there, early in 1930, that Tombaugh made photographic plates of the sky that confirmed Pluto's existence.

Mr. and Mrs. M.B. Tombaugh, parents of Clyde Tombaugh, are shown holding their other son Robert at their home in Burdett, Kansas, April 17, 1930. (AP Photo/Charles Grumich)

It was the discovery of a lifetime for Tombaugh and launched him on a long and productive career in astronomy.

Tombaugh, 74, now is professor emeritus of astronomy at New Mexico State University. He spoke with reporters in Chicago on Monday to mark the golden anniversary of his discovery.

Even now, little has been learned about tiny Pluto. Located some 2.7 billion miles from the sun, it meanders the frigid frontier of the solar system, taking 247 Earth years to complete one orbit of the sun. It is the ninth and outermost planet. But its lopsided orbit has temporarily swung it closer to the sun than Neptune.

In 1978, discovery of a moon of Pluto helped astronomers calculate that the planet has a mass of about one-fifth that of Earth's moon and a density comparable to water, leading many astronomers to suggest that Pluto may resemble a giant methane snowball.

The search for Pluto was started early in the 20th century by Dr. Percival Lowell, an American astronomer who had calculated the path of the planet Neptune. But he found slight irregularities in the predicted orbit of Neptune -- perturbations that Lowell decided could only be caused by the gravitational pull of an undiscovered planet.

Lowell started work at the Flagstaff observatory in 1906 to find his "Planet X" by photographing and comparing identical sky sections.

The theory behind his experiment was simple. A star appears relatively fixed in the night sky. But in comparison, a planet -- its very name means "wanderer" -- rushes along its orbital path, shifting drastically against the frozen backdrop of stars.

Lowell looked for the planet from 1905 until his death in 1916.

The search resumed in 1929, this time using two new pieces of equipment -- a 13-inch photographic telescope, and a blink comparator -- a devise that rapidly interchanged the photographs to be compared. If a heavenly body had moved appreciably, its image would appear and disappear as the photographs were rapidly switched.

Sitting for hours at a time on a mountaintop in a dark dome peering into the sky was tedious, painstaking work, too boring for someone with a doctor's degree in astronomy. So, directors of the observatory sought a highly skilled amateur to make the plates. They chose Tombaugh.

Clyde Tombaugh, Kansas farm boy who discovered the planet Pluto, 1931. (AP Photo)

Tombaugh said he had no idea when he took the job that he would end up exploring the distant fringes of the solar system.

"It didn't matter what they wanted me to do. Just anything to get off the farm. We got hailed out that year and were absolutely broke. When I got on that train to go to Flagstaff, I didn't have enough money for the return fare."

The original plan called for Tombaugh to make and develop the plates. An experienced astronomer would then perform the tedious task of comparing them using the comparator.

Soon, Tombaugh proved to be such a capable astronomer that he was asked to compare the plates as well. "My heart sank to my knees," Tombaugh said. Since he had made the plates, Tombaugh knew what a "starry mess" was pictured on them. Each image would have to be carefully examined. "I wasn't a bit happy," Tombaugh said of the task which made him famous.

Clyde Tombaugh poses with the telescope through which he discovered the planet Pluto at the Lowell Observatory on Observatory Hill in Flagstaff, 1931. (AP Photo)

In February 1930, after examining some 400,000 stars of the constellations of Taurus and the western portion of Gemini, Tombaugh was examining plates photographed in late January of the eastern part of the constellation. He had completed about two-thirds of the photographic field.

Suddenly, there it was.

A dim object was popping in and out of the background field of stars.

"That's it!" Tombaugh remembers exclaiming. "Everything was confirmed."

The observatory director held off telling the world until further tests confirmed the discovery.

Then on March 13, 1930, the 75th anniversary of Percival Lowell's birth, the world finally learned that Tombaugh had found the mystery planet that Lowell had predicted some 30 years before.

PLUTONIUM AND PLUTO
Albuquerque, New Mexico, Monday, June 10, 1991

When two men who changed the world more than half a century ago met for the first time, there was no mistaking them.

Glenn Seaborg, co-discoverer of plutonium 50 years ago, was the one with the periodic table of elements on his tie.

Dr. Glenn Seaborg, the Nobel Prize-winning physicist who discovered plutonium and has the Sg element on the elements table, Seaborgium, named after him, is shown at the University of California, Berkeley with an elements table sculpted by a fan. The Sg element is in the lower right corner, April 17, 1997. (AP Photo/Susan Ragan)

And Clyde Tombaugh, who spotted the planet Pluto in 1930, was the one wearing the Pluto watch - Pluto the cartoon dog, that is.

"This watch has no hands, it has paws," Tombaugh joked.

Visitors kiss a bust of Clyde Tombaugh at Lowell Observatory in Flagstaff, July 14, 2015. (AP Photo/Felicia Fonseca)

Seaborg, 79, and Tombaugh, 85, met Monday at Sandia National Laboratories after Seaborg gave a 50th anniversary speech about the discovery of plutonium and subsequent transuranic elements (those with higher atomic numbers than uranium).

Although the mood was light and the jokes plentiful, the two men were serious when they talked with reporters about their accomplishments.

"Plutonium plays such an important role in the affairs of man," said Seaborg, a Nobel Prize-winning chemist.

The discovery of the heavy, radioactive, man-made element led to the creation of the atomic bomb that was dropped on Japan to bring an end to World War II. It also made it possible for nuclear reactors to be used to produce electricity.

Tombaugh, the last person to discover a planet, said the question people most often ask him is whether there's a 10th planet in our solar system.

"I don't think the prospects are very good," he said. "There are no more planets out there."

Tombaugh, who lives in Mesilla Park, New Mexico, and is a professor emeritus at New Mexico State University, discovered the ninth planet while working at Lowell Observatory in Flagstaff, Arizona.

As Seaborg spoke to a Sandia labs audience, he mixed the complexity of the search for transuranic elements with a dash of humor and sprinkle of his personal life.

He recalled the day he announced that two transuranic elements had been discovered - and it wasn't at a Northwestern University chemistry conference November 16, 1945, as planned.

It was a week earlier, when he was a guest on the television show "Quiz Kids" and a child asked him a simple question.

"One of them asked, 'By the way, have any new elements been discovered?'" he said. "I blurted out 'Yes! Elements with the atomic numbers 95 and 96,' and that was the announcement to the world."

Seaborg, a professor at the University of California-Berkeley and an author, joked about how he and his colleagues arrived at the name and atomic symbol for plutonium.

He said the symbol should be Pl, not Pu, "but we just liked the roll of the P-U better."

They decided to name it after the most recently discovered planet, in the spirit of uranium and neptunium, two previous elements on the chart named after Uranus and Neptune.

But the group first considered names such as extremium and ultimium on the basis that the element they discovered must be the heaviest possible to synthesize, he said.

"Can you imagine how foolish we would've looked?" he said.

He has since co-discovered nine more transuranium elements such as americium, berkelium, californium, einsteinium, fermium and nobelium.

Seaborg said 109 elements have been discovered, and he continues to search for heavier elements "because it's fun."

"It just basically leads to a better understanding of atomic structure and a better understanding of nuclear structure, which is basic to everything," he said.

LOWELL OBSERVATORY
Tucson, Arizona, Monday, June 6, 1994

Percival Lowell, wealthy scion of a Massachusetts textile family, was obsessed with Mars: He sought to prove that life existed on the Red Planet.

So he dispatched a Harvard astronomer on a hasty rail tour to scour the Arizona Territory for a telescope site.

Lowell ultimately failed in his quest. But, 100 years after he first peered through a telescope from Mars Hill in Flagstaff on May 28, 1894, the observatory he founded is thriving.

It was from Lowell Observatory in the early part of the century that astronomers discovered the planet Pluto and gleaned the first clue that the universe is expanding.

The Lowell Observatory in Flagstaff, April 17, 1930. (AP Photo)

The Pluto Dome on the grounds of the Lowell Observatory, February 9, 2005. (AP Photo/Matt York)

The observatory, small by today's standards and not part of any university, continues its scientific role. It seeks out comets and asteroids, studies the planets and assesses the sun's long-term future and stability.

"They have a long and distinguished history in making contributions, particularly in planetary sciences, and they continue to do so," says Richard Green, director of the National Optical Astronomy Observatories, including Kitt Peak near Tucson.

The observatory marked its centennial by opening a $ 2.5 million interactive visitor center that will attract some of the thousands of people who pass through Flagstaff, 140 miles north of Phoenix, on their way to the Grand Canyon.

"The observatory is alive and well and pursuing astronomy at the forefront of what's being done," says Edward Bowell, a staff astronomer and noted asteroid hunter.

Bowell and astronomers Eugene and Carolyn Shoemaker head an effort to detect potentially lethal chunks of ice and rock hurtling through space.

They are among the most prolific patrollers in the world: Bowell has discovered and named more than 300 asteroids, and Carolyn Shoemaker has found more than 300 asteroids and 30 comets.

They troll for killer comets and asteroids, like Comet Shoemaker-Levy 9, which is expected to plunge into Jupiter on July 16 at 134,000 mph.

From left, Amateur astronomer David Levy, Dr. Eugene Shoemaker of the U.S. Geological Survey and Dr. Carolyn Shoemaker of the Lowell Observatory, display images of the comet Shoemaker-Levy 9 at a news conference at the Goddard Space Flight Center in Greenbelt, Maryland. The Shoemaker's and Levy co-discovered the comet, July 17, 1994. (AP Photo)

Shoemaker-Levy 9's collision with Jupiter will unleash a blast conservatively estimated at 1 million megatons, 200 times the explosive power of the world's entire nuclear arsenal.

Lowell, with 16 Ph.D. astronomers, has stayed in the game partly by upgrading older, undersized equipment in an era when instruments are constantly getting bigger and more sophisticated.

Most important has been the use of extremely sensitive, digital computer cameras that improve the performance of existing telescopes, according to Lowell director Robert Millis, co-discoverer in 1977 of rings around Uranus and in 1988 of Pluto's hazy atmosphere.

These devices allow an instrument the size of the 72-inch Perkins telescope, the largest at Lowell, to operate as well as the 200-inch Hale telescope on California's Mount Palomar did before its operation, too, was improved in the same way.

Kevin Schindler of Lowell Observatory stands next to the Pluto telescope, March 17, 2015. (AP Photo/Felicia Fonseca)

With an annual budget of about $ 2.5 million, Lowell operates four small telescopes at Mars Hill, elevation 7,200 feet, and has five more instruments operating or planned on 7,200-foot Anderson Mesa south of Flagstaff and one in Australia.

On Anderson Mesa are the Perkins scope, owned by Ohio State and Ohio Wesleyan universities, the 42-inch John S. Hall telescope, a 31-inch reflecting telescope, a $ 17 million instrument under construction called the Navy Prototype Optical Interferometer and a wide-field telescope recently refurbished and moved from Ohio to study approaching comets and asteroids.

The interferometer, to be operated in collaboration with the Naval Research Laboratory and the U.S. Naval Observatory, will use six 20-inch mirrors acting in concert to produce images 100 times sharper than conventional ground telescopes, Millis said.

Lowell, the world's largest privately owned institution of its kind, reflects the course charted by its independent, charismatic founder. Percival Lowell, who died in 1916, stipulated in its endowment that it couldn't become subsidiary to an educational institution.

Until 1948, research languished, since the observatory existed solely on interest from its endowment. Then trustee Roger Lowell Putnam sought other funding and landed a contract with the National Weather Service for a project to study the sun's stability that continues today.

The observatory was born out of Lowell's fascination with the work of Italian astronomer Giovanni Schiaparelli, who described lines on Mars' surface as "canali." The word literally means "grooves," but was mistakenly translated into English as "canals."

In early 1894, Lowell sent Harvard-trained astronomer A.E. Douglass to Arizona to scout a site from which to observe Mars. In six weeks, Douglass ruled out Tombstone, Tucson, Tempe and Prescott and settled on "Site 11," west of Flagstaff. Five weeks later, he had a wooden dome built for Lowell's two borrowed telescopes.

Lowell, a widely traveled Harvard intellectual, theorized that Schiaparelli's canali were farm irrigation systems. Lowell wrote three books about Mars, though professional astronomers rejected his views.

But later work at the observatory won acclaim.

The most notable finding was astronomer Vesto Melvin Slipher's 1912 discovery that the Andromeda nebula was moving three times faster than anything known. Slipher provided the first data used by Edwin Hubble to develop his theory that the universe is expanding, said Henry Giclas, 84-year-old Lowell astronomer emeritus.

Dr. Earl C. Slipher, Director of Lowell displays photos of Mars in Flagstaff, where the international Mars committee makes its headquarters. Dr. Slipher says that it may be 50 years before man is ready for a flight to Mars, February 25, 1958. (AP Photo)

Percival Lowell began a search for "Planet X," near Neptune, in 1902. He never found it. But the observatory's Clyde Tombaugh did in 1930, discovering Pluto - only 6 degrees from where Lowell calculated it would be.

TOMBAUGH DEAD AT 90

Las Cruces, New Mexico, Monday, January 18, 1997

Clyde Tombaugh, the astronomer who discovered the planet Pluto before he even had a college degree, is dead at the age of 90.

Clyde Tombaugh, 1990. (AP Photo/Will Yurman)

Tombaugh, who was an astronomy professor at New Mexico State University and founder of the school's research astronomy department, died Friday at his home in Mesilla Park, New Mexico.

"He was truly one of the great men of science," said university astronomer Jack Burns, a longtime friend.

Burns said Tombaugh had been struggling with breathing problems for the past two years.

Tombaugh discovered Pluto, the ninth planet from the sun, in 1930 when he had just turned 24, a Kansas farm boy who didn't yet have a college degree.

On the basis of his informal studies of Mars and Jupiter, the self-taught young astronomer had been given a job at the Lowell Observatory in Flagstaff, Arizona, helping in the search for what was then known only as Planet X.

Astronomer Percival Lowell had postulated the existence of Planet X because of wobbles in the orbits of Uranus and Neptune, which could be caused by the gravitational effects of an unseen planet.

For 10 months, Tombaugh compared 14- by 17-inch photos of the night sky, switching between two images taken several days apart. If a planet was visible among the images of stars, there would be a noticeable change in its position from one photo to the next.

"It was tedious," Tombaugh told The Associated Press in 1990. "But it was much more interesting work than farming as far as I was concerned."

On February 18, 1930, he spotted a small shift in the position of one object - the mysterious Planet X, later named Pluto.

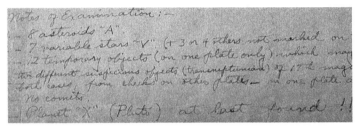

The original hand written notes by Clyde Tombaugh read "Planet "X" (Pluto) at last found!!!" on the aged paper that wraps around the original photographic plates of the planet Pluto at the Lowell Observatory in Flagstaff, February 9, 2005. (AP Photo/Matt York)

"For three-quarters of an hour, I was the only person in the world who knew exactly where Pluto was," he later said.

His discovery earned him a full scholarship to study astronomy at the University of Kansas. He came to New Mexico State University in 1955 and retired in 1973.

Tombaugh reacted with disdain last year when some scientists speculated that, based on photographs from the Hubble Space Telescope, Pluto isn't a planet at all but rather part of the comet belt on the edge of the solar system.

He called the claims "a bunch of nonsense."

Tombaugh is survived by his wife, Patsy, two children and several grandchildren and great-grandchildren.

Chapter 2

MYSTERIOUS PLUTO

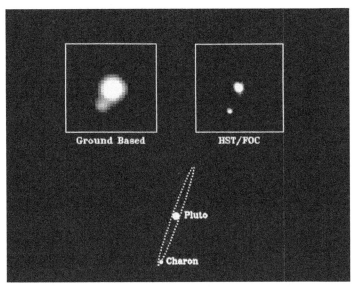

Ground Based   HST/FOC

Pluto

Charon

NASA's Hubble Space Telescope has obtained the clearest pictures ever of our solar system's most distant and enigmatic object, the planet Pluto, using the European Space Agency's Faint Object Camera. A recent Faint Object Camera image of Pluto and its satellite Charon is shown in the upper right-hand frame of this picture. At upper left is a photo of Pluto and Charon taken through an earth-based telescope, October 1, 1990. (AP photo)

PLUTO PROFILE
Monday, February 8, 1993

Facts about Pluto and its moon, Charon, (pronounced like the woman's name, Sharon).

Discovery
Astronomer Clyde Tombaugh discovered Pluto in 1930 from Lowell Observatory in Flagstaff, Arizona. Astronomer James

Christy found Charon in 1978, using the U.S. Naval Observatory near Flagstaff.

Origins

Some astronomers believe a collision of larger objects created Pluto and Charon. Another common theory, that Pluto was a moon of Neptune that escaped, has been challenged.

Size

Pluto's estimated diameter is 1,430 to 1,510 miles, the smallest known planet in our solar system. Charon's diameter is estimated at 746 to 795 miles - small, but in this solar system, the largest moon relative to the planet it orbits. For this reason, some astronomers call Pluto and Charon a "double planet."

Calendar

One year on Pluto equals 248 Earth years, the time it takes Pluto to complete one orbit around the sun. One month or one day on Pluto both equal 6.4 Earth days. That's because Pluto takes 6.4 Earth days to spin on its axis while Charon circles Pluto in the same time. No other planet has equivalent months and days.

Orbit

Pluto usually is the ninth and outermost planet. But its elliptical orbit brings it as near as 2.7 billion miles from the sun and as far as 4.6 billion miles, so sometimes Pluto is closer to the sun than Neptune. That's true between 1979 and 1999.Charon is 12,400 miles from Pluto.

Conditions

Cold and somewhat dark. Pluto's surface temperature is around 387 degrees below zero Fahrenheit. Sunlight hitting Pluto is 900 times less intense than that reaching Earth, but provides 250 times the illumination Earth gets from a full moon. Pluto appears to have ice caps, especially its bright south pole.

Gravity

Pluto's gravity is 1 percent to 2 percent that of Earth's. A 200-pound man on Earth would weigh 2 pounds to 4 pounds on Pluto.

Composition

Scientists believe Pluto is about three parts rock and one part frozen gases. Its icy surface is 97 percent frozen nitrogen; the rest is mostly frozen carbon monoxide and methane, or natural gas. Charon's surface contains water ice.

Atmosphere

Pluto's thin atmosphere contains methane but is otherwise a mystery. As Pluto moves away from the sun, its atmosphere freezes and falls to the planet's surface. By the year 2020, most of the atmosphere will be frozen and stay that way for about two centuries until Pluto again approaches the sun.

NOT ALWAYS THE MOST DISTANT PLANET
Washington, Sunday, January 21, 1979

For nearly 40 years, students have been learning that Pluto is the planet most distant from the sun. Starting Monday, that will no longer be true.

The U.S. Naval Observatory reports that on that day Pluto will edge inside the orbit of Neptune, making the planet named for the Greek sea god the most distant.

And that will remain the case until March 4, 1999, when the orbits of the two cross again and Pluto regains its last place ranking.

Since Pluto was only discovered in 1930 by Clyde Tombaugh at the Lowell Observatory, this is the first time scientists will actually be aware of it moving inside of Neptune.

However, they have calculated that this event takes place every 248 years, meaning that it last happened in 1731 and before that in 1483, 1235 and so forth, remaining inside the orbit of Neptune each time for 20 years.

The Naval Observatory reports that Pluto actually crossed Neptune's orbit last November, but since the two planets do not have circular orbits it will not be closer to the sun than Neptune until Monday. The time when this will occur cannot be calculated exactly.

Although the paths of the planets cross, there is no reason for alarm. They will never collide, the observatory added. Their orbits are so inclined that Pluto and Neptune can come no closer than about 240million miles to one another.

Only last year, scientists studying photographs of Pluto discovered that it appears to have a large satellite which would be visible only from one side of the planet. This moon has been named Charon.

Besides being most distant - most of the time - Pluto is the smallest known planet, with a diameter of about 3,000 kilometers, and a mass about one eighth that of our Moon.

Scientists believe the surface of Pluto is covered with frozen methane, and surmise that the planet is a ball of ice with water, methane and ammonia the major constituents.

Neptune, also with a large moon, was "discovered" in 1846, although it had been seen before that and mistaken for a star.

DOES CHARON EXIST?
Cambridge, Massachusetts, Thursday, April 17, 1980

An astronomical observation made in South Africa, if correct, bolsters the theory that the planet Pluto has a huge satellite, an astronomer said Thursday.

"The observation I suppose helps a little bit," said Dr. Brian Marsden of the Smithsonian Astrophysical Observatory. "We need more observations. I hope we get other reports."

Marsden was commenting on a report that Alistair R. Walker, an astronomer at the South African Astronomical Observatory in Sutherland, observed April 6 that the light from a star was blocked for about 50 seconds.

Scientists believe that the light may have been blocked by a body called Charon.

Charon, whose very existence is in question, was believed to have been first observed two years ago. It may be a satellite of Pluto, which is usually the most distant planet in the solar system. Pluto's orbit is approximately 3.6 billion miles from the Earth's orbit.

But Marsden said the body believed to be Charon may actually be Pluto.

"I'm 90 percent sure there is some kind of satellite there," Marsden said, basing that on the combined evidence, not just the South African observation.

Learning why Pluto has a satellite, if it does, might help astronomers understand the origin of the planet and the solar system, Marsden said.

Walker's observation, if correct, would indicate that Charon is about half the size of Pluto.

This would make Pluto-Charon "much more like a double planet, an extreme case where the satellite is very large in relation to the primary," Marsden said. The Earth, for example, is four times the size of its satellite, the moon.

But, he cautioned, "There may be other reasons why the star went out for those 50 seconds." He mentioned clouds as a possibility.

PLUTO HAS AN ATMOSPHERE
Tucson, Arizona, Thursday, October 6, 1980

Scientists have detected a thin atmosphere of methane on the distant planet Pluto, and they say the discovery discredits the theory that Pluto is not really a planet at all, but some other type of body, such as a comet.

"We feel that the recent discovery of (Pluto's) satellite and the present detection of an atmosphere will do much to enhance its image and establish Pluto as a more regular and respectable member of the planetary community," said Uwe Fink of the University of Arizona.

Pluto's moon was discovered in 1978.

Some scientists have speculated that Pluto was a comet or an asteroid, or even a runaway satellite of Neptune.

Fink, an associate professor with the university's Lunar and Planetary Laboratory, disclosed the discovery of an atmosphere Wednesday during the annual meeting of the Division for Planetary Sciences of the American Astronomical Society in Tucson.

"Pluto is the last planet for which the presence or absence of an atmosphere was in question," Fink said.

The tiny planet's atmosphere is composed of methane gas and is about 1-300th as dense as the total Earth atmosphere, he said. Current data suggests the presence of no other gases in Pluto's atmosphere.

"It is the least dense atmosphere in the solar system, but it's not too different from the atmosphere of Mars," Fink said.

Researchers discovered the atmosphere after observations in May with a special light detector attached to a spectrometer designed and built by the lab. The spectrometer, which measures light waves, was mounted on the university's 61-inch telescope.

Solar system planets: Mercury, Venus, Mars, Earth, Jupiter, Saturn, Neptune, Uranus and Pluto (l-r), (AP composite photo)

Scientists were looking for methane on Pluto because atmospheres of "all the major planets contain a lot of methane. So if there was an atmosphere on Pluto, then methane was a likely candidate.

"There's the possibility that the atmosphere exists only when Pluto is closest to the sun and then freezes out to a surface methane frost when Pluto is farther away," Fink said.

But because Pluto takes 248 years to revolve once around the sun, it won't be far enough away for astronomers to check the hypothesis through observations for 100 years, he said.

Pluto's orbit brings it to within about 2.8 billion miles of the sun at its closest point and 4.6 billion at its farthest. Earth orbits at about 93 million miles from the sun.

Of the nine known planets, Pluto is usually the farthest from the sun, although its irregular orbit sometimes brings it in closer than Neptune.

YES, THERE IS A CHARON
Austin, Texas, Thursday, February 21, 1985

An astronomy student has made the first recorded observation of the eclipse of the planet Pluto by its moon, Charon, proving the existence of the satellite, University of Texas officials say.

The brief sighting was made by Richard Binzel, a graduate student, early Sunday, at the university's McDonald Observatory in West Texas, school officials said Tuesday.

Scientists have said since 1978 that Pluto's moon, Charon, has existed but they had no visual evidence.

"Any astronomer you talk to would agree that the satellite does exist, but the International Astronomical Union has not officially recognized it because a direct photographic image of the satellite has yet to be obtained," Binzel said.

Binzel said he had been looking for eclipses since 1982.

"Pluto is very far away and the orbits of Pluto and the satellite are relatively close, so even through the largest ground-based telescopes, it is extremely difficult to detect the satellite," Binzel said.

Astronomers at Kitt Peak, Arizona, Mount Palomar, Calif., and Mauna Kea, Hawaii, also have been searching for the eclipse, he said.

Binzel said he coordinated his observation schedule Sunday night with David Tholen, an astronomer at the University of Hawaii, so that they could monitor Pluto continuously for an eight-hour period.

However, the eclipse occurred only during the time when the planet could be viewed from Texas.

The total amount of light emanating from Pluto and Charon dropped four percent as measured with a photometer attached to the McDonald Observatory's 36-inch telescope, Binzel said.

The sighting of the eclipse will help researchers determine the size of Pluto and Charon, but that calculation remains to be determined, he added.

CLOSEST TO EARTH IN OVER 200 YEARS
Austin, Texas, Tuesday, January 12, 1988

Astronomers are mounting a major effort to unlock the secrets of Pluto, a mysterious distant planet that is shrouded by methane snow and chilled by summertime temperatures of minus 378 degrees.

Astronomers said Tuesday that Pluto remains the least understood of the sun's planets, although its mysteries may yield in the coming months as the small planet makes its closest approach to the telescopes of Earth in more than two centuries.

Pluto, the most distant of the sun's planets, swings to within 2.8 billion miles of Earth during 1988 and 1989, its closest visit since 1740. During that time the planet and Charon, its only known moon, will eclipse each other while astronomers watch.

"This is one case where Nature was kind to us in that we didn't have to wait more than a century for the eclipses," said Richard P. Binzel of the Planetary Science Institute in Tucson.

Binzel, speaking at a meeting of the American Astronomical Society, said that during the eclipses astronomers should be able to confirm the size of Pluto and Charon, learn something about their composition and even gather hints about the surface features of the small planet.

"We know Pluto and Charon are very different from each other," he said. "Charon is a much darker body and Pluto has a reddish tint."

Pluto was discovered in 1930, but the planet is so far away and so small that little was learned about it for decades. Pluto studies intensified while its 248-year orbit carried it closer to Earth. Astronomers didn't discover Charon, which orbits the planet every 6 days, until 1978.

Since then, astronomers have discovered that Pluto appears to be covered with a methane snow that is now melting and evaporating slightly as the planet approaches its closest point to the sun.

"The methane atmosphere is expanding now and is very extensive," said Laurence Trafton of the University of Texas.

Trafton said Pluto is so small that its gravity is not strong enough to hold the methane vapor and molecules of the gas are shooting out into space. Some of the gas is smashing into Charon at speeds of more than 700 mph.

"This may be an example of a double planet system where the atmosphere of the two bodies interact with one another," said Trafton.

Charon orbits only 12,000 miles above Pluto, in contrast to the 240,000 miles separating the Earth and its moon. Also, said Trafton, the total mass of both Pluto and Charon is only 20 percent of the mass of Earth's moon. Pluto is only twice as large as its moon.

Alan Stern of the University of Colorado said he and Trafton plan to conduct a careful search for other moons in orbit around Pluto because astronomers are puzzled why none have been seen.

"In the outer solar system, most of the planets have lots of moons," said Stern. "Uranus has about a dozen and Jupiter has several dozen."

Stern said telescopes in Colorado and Texas will be used in the search for other moons, and predicted that "we'll be able to find one or to determine that there aren't any out there."

Binzel said the studies during Pluto's close approach also should be able to pick out specific surface features. He said scientists already know that dark spots scar the small planet's surface.

Stern said astronomers feel a particular obligation to study Pluto.

"Because no spacecraft has gone to Pluto, it is the last of the astronomers' planets," he said.

More than 1,000 astronomers are attending the American Astronomical Society meeting, which continues through Thursday.

HUBBLE GIVES PLUTO A CLOSER LOOK
Washington, Friday, March 8, 1996

New telescopic images of Pluto show that the distant planet is a place of frozen gases, icy polar caps and clusters of bright and dark features that astronomers can't explain.

The Hubble Space Telescope photos, in which Pluto resembles a fuzzy soccer ball, have excited astronomers who say they show the planet is a place of surprises and mysterious processes.

"We can see from the images that Pluto is the most variegated, contrasty object in the outer solar system," said Alan Stern, an astronomer at the Southwest Research Institute in Boulder, Colorado.

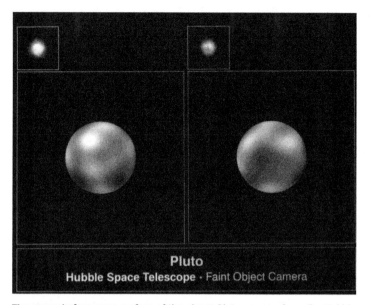

The never-before-seen surface of the planet Pluto as seen from the Hubble Space Telescope's Faint Object Camera. The two smaller inset pictures at top are the actual Hubble images. Opposite hemispheres of Pluto are seen in the bottom images, which are from a global map constructed through computer image processing of the Hubble data. The picture was taken when Pluto was 3 billion miles from Earth, March 6, 1996. (AP Photo/NASA)

He said the images confirm the belief of experts that Pluto is one of the strangest of the nine planets orbiting the sun.

Bright patches of white on Pluto, a frozen world 29 times farther from the sun than is the Earth, are probably fields of frozen nitrogen, while the dark patches may be hydrocarbons from the chemical splitting and freezing of methane, Stern said.

"Pluto never fails to surprise," said Bruce Margon, an astronomy professor at the University of Washington.

"It is like an icy little dwarf on the outskirts of the solar system," said Anne L. Kinney of the Space Telescope Institute in Baltimore.

Discovered just 66 years ago, Pluto was the last of the solar system planets to be identified. It's moon, Charon, was found in 1978, and it was not until 1988 that scientists discovered Pluto had an atmosphere.

Pluto is so strange that some astronomers have argued that it isn't really a planet, but some other object, such as an asteroid.

"It is a planet," said Stern. "It is round, it has a satellite (a moon), it has an atmosphere. There's only a minority view that it is not a planet."

Pluto is turned on its side, in relation to its orbit. This is the result, astronomers believe, of a gigantic collision with a comet or an asteroid that caused the small planet to tilt.

The planet is about two-thirds the size of the Earth's moon and is so far from the sun that sunlight on its surface is about as intense as moonlight is on Earth. Temperatures on Pluto drop to about minus 380 degrees Fahrenheit when the planet is at its farthest from the sun. Pluto made its closest approach to the sun in 1989 and temperatures rose to a balmy minus-350.

Stern said the slight thaw caused some nitrogen ice to evaporate into the atmosphere. He said that over the next few years Pluto will again cool and the gases will "snow" back on the surface, perhaps covering some the dark spots seen in the Hubble photos.

Before the photos by the Earth-orbiting Hubble, Pluto had been seen as just a tiny spot of light faintly detected, 4 billion miles from Earth.

The new pictures, which astronomers said pushed the Hubble to the very limit of its ability, aren't good enough to determine if Pluto has mountains and valleys like those on Earth and Mars.

Pluto takes 248 years to circle the sun and during all but 20 of those years it is the farthest out of all the planets. It crossed inside the orbit of Neptune in 1979. In three years, it will again move out to resume its most-distant place.

A robot craft, called the Pluto Express, is scheduled to be launched toward Pluto in the next decade, but it will take 10 to 12 years for the craft to reach the planet.

BEYOND OUR PLANETS
New York, Thursday, June 5, 1997

Astronomers have found an icy miniplanet that orbits the sun well beyond Pluto, providing evidence that the solar system extends much farther than was once thought.

The little planet is about 300 miles across, which gives it a surface area comparable to Texas. It is the brightest solar system object to be found beyond Neptune since the discovery of Pluto's moon Charon in 1978.

At its most distant, it wanders three times farther from the sun than Pluto, tracing a looping, oblong path into an astronomical terra incognita.

"It's the first object in a sort of no man's land, an area we never thought we could get a glimpse of with our current technology," said Jane Luu, an astronomer at the Harvard-Smithsonian Center for Astrophysics in Cambridge, Mass.

And it's probably not alone. Theoretical calculations suggest that there are millions of small, icy solar system objects well beyond the outermost planets.

Astronomers consider their new discovery an extension of the Kuiper belt, a collection of small, icy bodies that circle the sun beyond the orbit of Neptune. About 40 Kuiper belt objects have been discovered since 1992.

Illustration from NASA's Hubble Space Telescope team showing largest known Kuiper Belt objects, (NASA via AP)

Before then, the only known Kuiper belt objects were the planet Pluto, discovered in 1930, and Charon.

Luu discovered the new object, known as 1996TL66, with colleagues from Harvard, the University of Hawaii and the University of Arizona, as well as an amateur astronomer based in Cloudcroft,

New Mexico. They describe the find in today's issue of the journal Nature.

"I wouldn't call this a major planet," said Brian Marsden, a Harvard astronomer and contributor to the Nature paper. "But then I tend not to call Pluto a major planet."

In fact, 1996TL66 is considered too minor to be named for a Roman god, like the other planets.

Astronomers surmise that it is composed of the same material as other outer solar system objects - water, carbon dioxide, methane and other materials - all frozen solid.

The astronomers found 1996TL66 with a University of Hawaii telescope as the object passed among the outer planets last October. They and others followed it for several months with telescopes in Hawaii, Arizona and New Mexico.

The object's motion over the last few months shows that it follows a lopsided orbit unlike that of any other Kuiper belt object. It swings through the neighborhood of the outermost planets every 800 years, then loops far out into space before making its next pass.

Astronomers have never seen such a thing.

"It just reminds us that we really don't know what the outer solar system holds," Luu said.

Some researchers had an inkling that the object would be out there, however. Hal Levison, a scientist at the Southwest Research Institute in Boulder, Colorado, has been telling his colleagues for the past year that objects such as 1996TL66 ought to exist.

The research behind that prediction, conducted with Martin Duncan of Queen's University in Canada, is scheduled for publication in a future issue of the journal Science.

"Though we knew that we had an interesting scientific result, it just never occurred to me that anybody would find one," Levison said. "It's very satisfying when someone confirms your theories."

In his Science paper, Levison describes how Uranus and Neptune probably generated the Kuiper Belt during the formation of the solar system more than 4 billion years ago. According to that theory, Neptune kicked a small percentage of the Kuiper belt objects into oblong orbits such as the one followed by 1996TL66.

Luu and her colleagues found the object at the very beginning of a systematic search for objects at the edge of the solar system. Because it was so easy to find, the researchers calculate that there are hundreds, and perhaps more than 1,000, objects similar to 1996TL66.

"Unless we are improbably lucky, it is merely the first detected of a larger population of similar bodies," the astronomers wrote.

BACK TO NORMAL
Washington, Thursday, February 11, 1999

Tiny Pluto slips outside the orbit of Neptune today to resume its role as the farthest planet from the sun.

This return to normal comes just days after the littlest planet survived an attack that threatened to strip it of its planetary status altogether.

Normally the most distant planet, Pluto has an unusual orbit that takes 248 Earth-years to complete one trip around the sun. During just 20 of those years, it moves inside Neptune's orbit to become the eighth planet instead of the ninth.

Pluto moved inside Neptune's orbit on February 7, 1979, and was on course to cross back outside at 5:08 a.m. EST today, scientists at the National Aeronautics and Space Administration calculated.

Pluto will remain the most distant planet for the next 228 years.

Just last week, the Paris-based International Astronomical Union, the world's leading astronomical organization, reaffirmed Pluto's standing as the smallest planet.

News reports had said Pluto might be demoted to a minor planet, or - worse - a trans-Neptunian object.

"No proposal to change the status of Pluto as the ninth planet in the solar system has been made by any division, commission or working group of the IAU responsible for solar system science," said the 80-year-old organization, the final authority on astronomical matters.

Even though Pluto was crossing Neptune's orbit, there was no worry about a collision, NASA said, because the planets were going to be far apart at the time.

Pluto was discovered February 18, 1930, by Clyde Tombaugh at Lowell Observatory in Flagstaff, Arizona Its moon, Charon, was found in 1979.

With a diameter of 1,430 miles, Pluto is less than half the size of any other planet and only two-thirds as big as Earth's moon.

Chapter 3

DEMOTED

Los Angeles-based Web programmer Chris Spurgeon works at a computer displaying a bumper sticker, "Honk if Pluto is still a planet," that took him 15 minutes to design, at his office in Los Angeles. Scores of Web-savvy sellers hoping to support, and cash in on, Pluto's demotion to a "dwarf planet" bombarded the Internet hawking Pluto memorabilia worthy of a presidential candidate, from T-shirts and mugs to bumper stickers and mouse pads, August 25, 2006. (AP Photo/Nick Ut)

MINOR LEAGUE PLANETS
Albuquerque, New Mexico, Monday, January 25, 1999

Pluto's days as one of our solar system's nine major planets may be numbered.

Two groups within the International Astronomical Union are considering reclassifying the tiny planet.

One group, responsible for cataloging the "minor planets," wants to give Pluto the distinction of being minor planet number 10,000.

Another group, meanwhile, is looking at designating Pluto the first of a new class of objects that resemble giant comets circling the sun beyond the orbit of Neptune.

Pluto, the smallest of the major planets and the ninth from the sun, was discovered in 1930 by Clyde Tombaugh, a young astronomer working at the Lowell Observatory in Flagstaff, Arizona.

Guided tour of the Lowell Observatory as it stops at the Pluto Discovery Telescope, inside the Pluto Dome, in Flagstaff, Arizona. Percival Lowell founded the observatory in 1894 to look at Mars and other objects, and it has been the site of some major astronomical discoveries, including the first evidence the universe is expanding, Pluto and the rings of Uranus, February 9, 2005. (AP Photo/Matt York)

Tombaugh, who died two years ago at the age of 90, came to New Mexico in 1955 and founded the astronomy department at New Mexico State University in Las Cruces, where he lived the rest of his life.

Pluto's stature as a major planet has long been questioned.

With a diameter of 1,440 miles - smaller than the Earth's moon - many astronomers argue that it's more like a big asteroid or a giant comet.

And while the other planets circle the sun with roughly circular orbits, Pluto's is a great sweeping ellipse that frequently takes it closer to the sun than Neptune, the eighth planet from the sun.

Brian Marsden of the International Astronomical Union's Minor Planet Center says Pluto should be designated as a minor planet.

The minor planets are identified by a number, and the tally recently stood at 9,913. The number 10,000 may come up soon because of new discoveries.

In the past, Marsden has saved the even multiples of a thousand - 1,000, 2,000 and so on - for objects worthy of special recognition.

And so he wants to give the designation "10,000" to Pluto.

It's a way of recognizing Pluto's importance, while also recognizing that it's not a major planet, Marsden said.

JUST A LUMP OF ICE?
New York, Friday, January 26, 2001

One of the nation's leading science museums has quietly shaken up the universe by suggesting that Pluto is not necessarily a planet at all but just a lump of ice.

The startling suggestion comes from scientists at the Rose Center for Earth and Space, which opened last year at the American Museum of Natural History in New York.

There is a 9-foot-diameter model of Jupiter hanging from the ceiling at the center. There is Saturn with its rings, Mercury, Venus, Earth, Mars, Neptune and Uranus. But what about Pluto, long considered the ninth planet in the solar system?

A solar system display says: "Beyond the outer planets is the Kuiper Belt of comets, a disk of small, icy worlds including Pluto."

"There is no scientific insight to be gained by counting planets," says Neil de Grasse Tyson, director of the Hayden Planetarium, the centerpiece of the Rose Center. "Eight or nine, the numbers don't matter."

Many astronomers say the museum, the first prominent institution to take this position, has overstepped its bounds.

"Tyson is so far off base with Pluto, it's like he's in a different universe," says David Levy, author of "Clyde Tombaugh, Discoverer of Planet Pluto," about the Kansas farm boy who first spotted Pluto. "The majority of astronomers have said that unless there is definitive evidence to the contrary, Pluto stays a major planet."

The International Astronomical Union calls Pluto one of nine planets in the solar system, and a 1999 proposal to list Pluto as both a planet and a member of the Kuiper Belt was abandoned after it drew strong opposition from astronomers who did not want to diminish Pluto's status.

Pluto has always been a little different: Its composition is like a comet's, and its elliptical orbit is tilted 17 degrees from the orbits of the other planets.

When Pluto was discovered in 1930, it was thought to be about the same size as Earth, but astronomers have now learned that it is only 1,413 miles wide - smaller than the Earth's moon.

Then, in 1992, astronomers discovered the first Kuiper Belt object, and since then have found hundreds of chunks of rock and ice beyond Neptune, including about 70 that share orbits similar to Pluto's.

The Rose Center says there is no universal definition of a planet and instead divides the solar system into the sun and five families of objects.

Tova Hagler,10, left, reads through the names of the planets with her brother Yaakov, 5, as they walk through the Scales of the Universe exhibit at the Rose Center for Earth and Space at the American Museum of Natural History in New York. The museum has sprouted a small controversy over no longer classifying Pluto as a planet, January 25, 2001. (AP Photo/Beth A. Keiser)

There are terrestrial planets, or small, dense rocky objects like Mercury, Venus, Earth and Mars; an asteroid belt consisting of craggy chunks of rock and iron between Mars and Jupiter; the gas giants, which are Jupiter, Saturn, Uranus and Neptune; and two reservoirs of comets, the Oort Cloud and the Kuiper Belt. And Pluto?

"It's in the Kuiper Belt," Tyson says. "What's it made of? It's mostly ice."

Tyson says there is a precedent to demoting planets: Ceres was called a planet in 1801 and later demoted. Critics counter that Ceres, which is only 580 miles wide, was only considered a planet for a year, while Pluto has been a major planet for more than 70 years. In addition, they say, there was consensus among astronomers in the case of Ceres.

Still, others praise the museum for its bold move.

"People just don't like the idea that you can change the number of planets," says David Jewit, a professor at the University of Hawaii who co-discovered the first Kuiper Belt object. "The Rose Center is just slightly ahead of its time."

Jane Levenson, an "explainer" at the Rose center, says visitors - mostly kids - sometimes ask about the missing Pluto.

"We just explain that there are five types of objects that circle the sun," she says. "We don't make a big deal about Pluto."

IDENTITY CRISIS
Los Angeles, Monday, August 14, 2006

Our solar system is suffering an identity crisis.

For decades, it has consisted of nine planets, even as scientists debated whether Pluto really belonged. Then the recent discovery of an object larger and farther away than Pluto threatened to throw this slice of the cosmos into chaos.

Should this newly found icy rock known as "2003 UB313" become the 10th planet? Should Pluto be demoted? And what exactly is a planet, anyway?

Ancient cultures regularly revised their answer to the last question and present-day scientists aren't much better off; There still is no universal definition of "planet."

That all could soon change, and with it science textbooks around this planet.

At a 12-day conference beginning Monday, scientists will conduct a galactic census of sorts. Among the possibilities at the meeting of the International Astronomical Union in the Czech Republic capital of Prague: Subtract Pluto or christen one more planet, and possibly dozens more.

"It's time we have a definition," said Alan Stern, who heads the Colorado-based space science division of the Southwest Research Institute of San Antonio. "It's embarrassing to the public that we as astronomers don't have one."

The debate intensified last summer when astronomer Michael Brown of the California Institute of Technology announced the discovery of a celestial object larger than Pluto. Like Pluto, it is a member of the Kuiper Belt, a mysterious disc-shaped zone beyond Neptune containing thousands of comets and planetary objects. (Brown nicknamed his find "Xena" after a warrior heroine in a cheesy TV series; pending a formal name, it remains 2003 UB313.)

NASA image of Kuiper Belt object 2003 UB313 (nicknamed "Xena") and its satellite, Gabrielle, artist rendition (AP)

The Hubble Space Telescope measured the bright, rocky object at about 1,490 miles in diameter, roughly 70 miles longer than Pluto. At 9 billion miles from the sun, it is the farthest known object in the solar system.

The discovery stoked the planet debate that had been simmering since Pluto was spotted in 1930.

Some argue that if Pluto kept its crown, Xena should be the 10th planet by default it is, after all, bigger. Purists maintain that there are only eight traditional planets, and insist Pluto and Xena are poseurs.

"Life would be simpler if we went back to eight planets," said Brian Marsden, director of the astronomical union's Minor Planet Center in Cambridge, Mass.

Still others suggest a compromise that would divide planets into categories based on composition, similar to the way stars and galaxies are classified. Jupiter could be labeled a "gas giant planet," while Pluto and Xena could be "ice dwarf planets."

"Pluto is not worthy of being called just a plain planet," said Alan Boss, an astrophysicist at the Carnegie Institution in Washington, D.C. "But it's perfectly fine as an ice dwarf planet or a historical planet."

The number of recognized planets in the solar system has seesawed based on new findings. Ceres was initially classified as a planet in the 1800s, but was demoted to an asteroid when similar objects were found nearby.

This photo from a sequence of images provided by NASA, taken from the Dawn spacecraft of Ceres, a dwarf planet located in the asteroid belt between Mars and Jupiter, April 24, 2015. (NASA via AP)

Despite the lack of scientific consensus on what makes a planet, the current nine and Xena share common traits: They orbit the sun. Gravity is responsible for their round shape. And they were not formed by the same process that created stars.

Brown, Xena's discoverer, admits to being "agnostic" about what the international conference decides. He said he could live with eight planets, but is against sticking with the status quo and would feel a little guilty if Xena gained planethood because of the controversy surrounding Pluto.

"If UB313 is declared to be the 10th planet, I will always feel like it was a little bit of a fraud," Brown said.

For years, Pluto's inclusion in the solar system has been controversial. Astronomers thought it was the same size as Earth, but later found it was smaller than Earth's moon. Pluto is also odd in other ways: With its elongated orbit and funky orbital plane, it acts more like other Kuiper Belt objects than traditional planets.

Even so, Pluto remained No. 9 because it was the only known object in the Kuiper Belt at the time.

When new observations in the 1990s confirmed that the Kuiper Belt was sprinkled with numerous bodies similar to Pluto, some scientists piped up. In 1999, the international union took the unusual step of releasing a public statement denying rumors that the ninth rock from the sun might be kicked out.

That hasn't stopped groups from attacking Pluto's planethood. In 2000, the Hayden Planetarium at New York's American Museum of Natural History unleashed an uproar when it excluded Pluto as a planet in its solar system gallery.

Earlier this year, NASA's New Horizons spacecraft began a 9 1/2-year journey to Pluto on a mission that scientists hope will reveal more about the oddball object.

The trick for astronomers meeting in Prague is to set a criterion that makes sense scientifically. Should planets be grouped by location, size or another marker? If planets are defined by their size, should they be bigger than Pluto or another arbitrary size? The latter could expand the solar system to 23, 39 or even 53 planets.

It's not an academic exercise the public may not be open to a flood of new planets. Despite their differences, scientists agree any definition should be flexible enough to accommodate new discoveries.

"Science progresses," said Boss of the Carnegie Institution. "Science is not something that's engraved on a steel tablet never to be changed."

DEMOTED TO DWARF
Prague, Czech Republic, Thursday, August 24, 2006

For 76 years, Pluto has been a shady character loitering in the darkest, roughest corner of Earth's neighborhood. On Thursday, astronomers made the planet named for the Roman god of the underworld an offer it couldn't refuse.

Dramatically reversing course just a week after floating the idea of reaffirming Pluto's planetary status, the International Astronomical Union abruptly dumped it and renamed it a dwarf, shrinking the solar system to eight planets from the traditional nine.

Members of the International Astronomic Union (IAU), vote on a resolution for planet definition during the IAU 26th General Assembly in Prague, Czech Republic. The world's leading astronomers voted on Pluto being newly classified as a "Dwarf Planet", August 24, 2006. (AP Photo/Petr David Josek)

The decision came as leading astronomers for the first time since the ancient Greeks dubbed planets heavenly "wanderers" approved a new definition for what is and isn't a planet. Puny Pluto, a planet since 1930, was booted because its oblong orbit strays into Neptune's: a no-no under the new guidelines.

It's not personal, just business, said Richard Binzel, a professor of planetary science at the Massachusetts Institute of Technology who helped hammer out the new rules.

Outgoing International Astronomical Union (IAU) President Ron Ekers of Australia, left, incoming IAU President Catherine Cesarski of France, center, and head of the Planet Definition Committee Richard Binzel of USA attend the news conference closing the XXVI general assembly of IAU in Prague, August 24, 2006. (AP Photo/CTK, Stan Peska)

"This is really all about science, which is all about getting new facts," he said, holding a plush toy of Walt Disney's Pluto dog. "Science has marched on ... Many more Plutos wait to be discovered."

Astronomers have labored without a universal definition of a planet since well before the time of Copernicus, who proved the Earth revolves around the sun. The new rules say a planet not only must orbit the sun and be large enough to assume a nearly round shape, but must "clear the neighborhood around its orbit."

Pluto doesn't. And predictably, its sudden demotion provoked plenty of wistful nostalgia.

"It's disappointing in a way, and confusing," said Patricia Tombaugh, the 93-year-old widow of Pluto discoverer Clyde Tombaugh.

"I don't know just how you handle it. It kind of sounds like I just lost my job," she said from Las Cruces, New Mexico. "But I understand science is not something that just sits there. It goes on. Clyde finally said before he died, 'It's there. Whatever it is. It is there.'"

The decision by the IAU, the official arbiter of heavenly objects, restricts membership in the elite cosmic club to the eight classical planets: Mercury, Venus, Earth, Mars, Jupiter, Saturn, Uranus and Neptune.

"This decision in fact affects the way we look at the evolution of our solar system," said astrophysicist Michael Shara of the American Museum of Natural History.

Pluto and objects like it will be known as "dwarf planets," which raised some thorny questions about semantics: If a raincoat is still a coat, and a cell phone is still a phone, why isn't a dwarf planet still a planet?

NASA said Pluto's downgrade would not affect its $700 million New Horizons spacecraft mission, which this year began a 9 1/2-year journey to the oddball object to unearth more of its secrets.

But mission head Alan Stern said he was "embarrassed" by Pluto's undoing and predicted that Thursday's vote would not end the debate. Although 2,500 astronomers from 75 nations attended the conference, only about 300 showed up to vote.

"It's a sloppy definition. It's bad science," he said. "It ain't over."

The shift also poses a challenge to the world's teachers, who will have to scramble to alter lesson plans just as schools open for the fall term.

"We will adapt our teaching to explain the new categories," said Neil Crumpton, who teaches science at a high school north of London. "It will all take some explanation, but it is really just a reclassification and I can't see that it will cause any problems. Science is an evolving subject and always will be."

Under the new rules, two of the three objects that came tantalizingly close to planethood will join Pluto as dwarfs: the asteroid Ceres, which was a planet in the 1800s before it got demoted, and 2003 UB313, an icy object slightly larger than Pluto whose discoverer, Michael Brown of the California Institute of Technology, has nicknamed "Xena." The third object, Pluto's largest moon, Charon, isn't in line for any special designation.

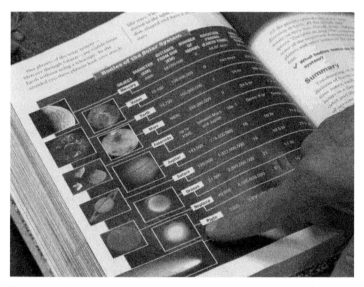

Tim Maggard, director of instructional technology for Hardin County, Kentucky schools, points to a reference in a textbook currently being used that still lists Pluto as a full-fledged planet, August 8, 2012. (AP Photo/Bruce Schreiner)

Brown, whose Xena find rekindled calls for Pluto's demise because it showed it isn't nearly as unique as it once seemed, waxed philosophical.

"Eight is enough," he said, jokingly adding: "I may go down in history as the guy who killed Pluto."

Demoting the icy orb named for the Roman god of the underworld isn't personal it's just business said Jack Horkheimer, director of the Miami Space Transit Planetarium and host of the PBS show "Star Gazer."

"It's like an amicable divorce," he said. "The legal status has changed but the person really hasn't. It's just single again."

AP Science Writers Alicia Chang in Los Angeles and Seth Borenstein in Washington, and correspondents Sue Leeman in London and Mike Schneider in Cape Canaveral, Florida., contributed to this report.

'BAD SCIENCE'
Los Angeles, Thursday, August 24, 2006

Astronomers rejiggered the nine-planet solar system for the first time in 76 years on Thursday, kicking out Pluto and creating a new category of "dwarf planets." Here are some questions and answers about our new solar system:

Q: Why is Pluto no longer a planet?

A: Pluto lost its planethood because it is not the dominant object in its region of space, according to new guidelines endorsed by the International Astronomical Union. Pluto's peculiar orbit overlaps with the much-larger Neptune's and it is one of thousands of icy bodies in the Kuiper Belt, a region in the outer solar system. Under the new definition, a planet must be a round object that orbits the sun and must also dominate its neighborhood.

Q: So how many planets do we now have in the solar system?

A: There are eight planets: Mercury, Venus, Earth, Mars, Jupiter, Saturn, Uranus and Neptune. Pluto has been demoted to a "dwarf planet."

Q: But didn't astronomers just last week propose keeping Pluto and adding three new planets?

A: Yes. A high-ranking panel of the IAU floated a proposal to expand the solar system and keep Pluto's status intact. It would have added the asteroid Ceres, Pluto's moon Charon, and a body larger than Pluto discovered last year, 2003 UB313. That plan also opened the door to adding even more planets. But after facing resistance from rank-and-file members who complained the proposal was too broad, the IAU went back to the drawing board and narrowed the definition.

Q: Are all scientists happy with the new eight-planet solar system?

A: No. Alan Stern of the Southwest Research Institute, who heads the New Horizons spacecraft mission to Pluto, thinks the new planet definition is ambiguous and called Pluto's demotion "bad science."

Q: What led astronomers to change the planet definition now?

A: Since the 1990s, scientists have discovered dozens of Pluto-like objects in the Kuiper Belt, all smaller than Pluto. But it wasn't until astronomer Michael Brown at the California Institute of Technology spotted the larger UB313 that scientists wrestled with what to call it. That forced them to rethink Pluto's status.

Q: Have all the planets been discovered?

A: Scientists say it's highly unlikely, but not impossible, that someone would find a ninth planet. Many objects that are still awaiting their designation would most likely qualify as dwarf planets. As Brown puts it: "It's going to be hard to find a new planet. You'll have to find something the size of Mars or larger. But boy, if I could find a planet, that'll be something."

HONK IF PLUTO IS A PLANET
Prague, Czech Republic, Thursday, August 24, 2006

It's smaller than Earth's moon, has a funky way of orbiting the sun, and lurks so far out on the fringes of the solar system even the powerful Hubble Space Telescope has to squint to see it.

Pluto is no stranger to controversy. In fact, it's been dogged by disputes ever since its discovery in 1930.

Many astronomers contend the ninth rock from the sun officially downgraded to "dwarf planet" status Thursday by the International Astronomical Union never deserved to be a full planet in the first place.

"They are just saying officially what we had been hinting to children anyway: that Pluto is very different from other planets," said Cedric Courson, program director in Paris for Planet Sciences, an association that promotes science at schools and among the public with support from the French Education Ministry.

Discovered by Clyde Tombaugh of Arizona's Lowell Observatory, Pluto was classified as a planet because scientists initially believed it was the same size as Earth. It remained one because for years, it was the only known object in the Kuiper Belt, an enigmatic zone beyond Neptune that's teeming with comets and other planetary objects.

Pluto got an ego boost in 1978 when it was found to have a moon that was later named Charon. The Hubble turned up two more, which this past June were christened Nix and Hydra.

But in the 1990s, more powerful telescopes revealed numerous bodies similar to Pluto in the neighborhood. New observations also showed that Pluto's orbit was oblong, sending it soaring well above and beyond the main plane of the solar system where Earth and the other seven planets circle the sun.

That prompted some galactic grumbling from astronomers who began openly attacking Pluto's planethood.

At one point, things looked so bad for Pluto, the international union said publicly in 1999 that rumors of Pluto's imminent demise were greatly exaggerated and there were no plans to kick it out of the cosmic club.

A year later, the Hayden Planetarium at New York's American Museum of Natural History was accused of snubbing Pluto by excluding it from a solar system exhibition.

Pluto took another hit after Michael Brown of the California Institute of Technology discovered 2003 UB313, a slightly larger Kuiper Belt object. What's the point, some astronomers wondered, in keeping Pluto as a planet?

"The public is not going to like the fact that Pluto has just been kicked out," Brown said Thursday. "But it's the right thing to do scientifically."

Los Angeles-based Web programmer Chris Spurgeon's bumper sticker, "Honk if Pluto is still a planet," August 25, 2006. (AP Photo/Nick Ut)

Ironically, Pluto's future had brightened earlier this year, when NASA sent the New Horizons spacecraft to Pluto to get a closer look at the ball of rock and ice. The Hubble has managed to glimpse only its most prominent surface features; New Horizons, if all goes well, will arrive in 2015.

As recently as last week, the IAU the official arbiter of heavenly bodies appeared ready to reaffirm Pluto's planet status.

Richard Binzel, a professor of planetary science at the Massachusetts Institute of Technology and a member of the international union's planet definition committee, had contended that Pluto met key tests of planetary physics "by a long shot" and had earned its status.

Before Thursday's vote, a weary-looking Binzel was asked whether he'll mourn Pluto.

"I don't know. Ask me later," he said, and walked away.

OUT OF THE COSMIC CLUB
Prague, Czech Republic, Thursday, August 24, 2006

For decades, it's been confused with a cartoon dog and ridiculed as a puny poser. Now Pluto, the solar system's consummate cling-on, has suffered its worst humiliation: It's not even a planet anymore.

Disney character Pluto in New York's Times Square, October 17, 2012. (AP Photo/Richard Drew)

After a tumultuous week of clashing over the essence of the cosmos, leading astronomers Thursday stripped Pluto of the planetary status it has held since its discovery in 1930. The new definition of what is and isn't a planet fills a centuries-old black hole for scientists who have labored since Copernicus without one.

The historic vote by the International Astronomical Union officially shrinks Earth's neighborhood from the traditional nine planets to eight.

But the scientists made clear they're as sentimental as anyone else about the ninth rock from the sun.

Jocelyn Bell Burnell a specialist in neutron stars from Northern Ireland who oversaw the proceedings in Prague urged those who might be "quite disappointed" to look on the bright side.

"It could be argued that we are creating an umbrella called 'planet' under which the dwarf planets exist," she said, drawing laughter by waving a stuffed Pluto of Walt Disney fame beneath a real umbrella. Later, she hugged the doll as she stood at the dais.

"Many more Plutos wait to be discovered," added Richard Binzel, a professor of planetary science at the Massachusetts Institute of Technology.

The decision by the prestigious international group spells out the basic tests that celestial objects will have to meet before they can be considered for admission to the elite cosmic club.

Members of the International Astronomic Union (IAU), vote on a resolution for planet definition during the IAU 26th General Assembly in Prague, Czech Republic. The world's leading astronomers voted on Pluto being newly classified and called as "Dwarf Planet", August 24, 2006. (AP Photo/Petr David Josek)

For now, membership will be restricted to the eight "classical" planets in the solar system: Mercury, Venus, Earth, Mars, Jupiter, Saturn, Uranus and Neptune.

Much-maligned Pluto named for the God of the underworld doesn't make the grade under the new rules for a planet: "a celestial body that is in orbit around the sun, has sufficient mass for its self-gravity to overcome rigid body forces so that it assumes a ... nearly round shape, and has cleared the neighborhood around its orbit."

Pluto is automatically disqualified because its oblong orbit overlaps with Neptune's.

Instead, it will be reclassified in a new category of "dwarf planets," similar to what long have been termed "minor planets." The definition also lays out a third class of lesser objects that orbit the sun "small solar system bodies," a term that will apply to numerous asteroids, comets and other natural satellites.

Experts said there could be dozens of dwarf planets catalogued across the solar system in the next few years handing the world's school teachers a challenge.

Neil Crumpton, a science teacher at Mountfitchet High School in Stansted Mountfitchet, north of London, called the announcement "very exciting."

"To be honest, this has been brewing for a while. Pluto has always been a bone of contention among astronomers because of the odd way it orbits the sun," Crumpton said. "For a start, we'll have to change all the mnemonics we use to teach children the lineup of the planets. But Pluto has not disappeared and it doesn't hurt children to know about it."

NASA said Thursday that Pluto's demotion would not affect its $700 million New Horizons spacecraft mission, which earlier this year began a 9 1/2-year journey to the oddball object to unearth more of its secrets.

"We will continue pursuing exploration of the most scientifically interesting objects in the solar system, regardless of how they are categorized," Paul Hertz, chief scientist for the science mission directorate, said in a statement.

The decision at a conference of 2,500 astronomers from 75 countries was a dramatic shift from just a week ago, when the group's leaders floated a proposal that would have reaffirmed

Pluto's planetary status and made planets of its largest moon and two other objects.

That plan proved highly unpopular, splitting astronomers into factions and triggering days of sometimes combative debate that led to Pluto's undoing. In the end, only about 300 astronomers cast ballots.

Now, two of the objects that at one point were cruising toward possible full-fledged planethood will join Pluto as dwarfs: the asteroid Ceres, which was a planet in the 1800s before it got demoted, and 2003 UB313, an icy object slightly larger than Pluto whose discoverer, Michael Brown of the California Institute of Technology, has nicknamed "Xena."

Charon, the largest of Pluto's three moons, is no longer under consideration for any special designation.

Brown was pleased by the decision. He had argued that Pluto and similar bodies didn't deserve planet status, saying that would "take the magic out of the solar system."

"UB313 is the largest dwarf planet. That's kind of cool," he said.

But as it all sank in, he added: "Deep down inside, I know this is the right thing to do. It's sad. As of today, I have no longer discovered a planet."

GIANT 9TH PLANET ON EDGE OF SOLAR SYSTEM?
Cape Canaveral, Florida, Wednesday, January 20, 2016

The solar system may have a ninth planet after all.

This one is 5,000 times bigger than outcast Pluto and billions of miles farther away, say scientists who presented "good evidence" for a long-hypothesized Planet X on Wednesday.

The gas giant is thought to be almost as big as its nearest planetary neighbor Neptune, quite possibly with rings and moons. It's so distant that it would take a mind-blowing 10,000 to 20,000 years to circle the sun.

Planet 9, as the pair of California Institute of Technology researchers calls it, hasn't been spotted yet. They base their prediction on mathematical and computer modeling, and anticipate its discovery via telescope within five years or less.

The two reported their research Wednesday in the Astronomical Journal because they want people to help them look for it.

"We could have stayed quiet and quietly spent the next five years searching the skies ourselves and hoping to find it. But I would rather somebody find it sooner, than me find it later," astronomer Mike Brown told The Associated Press.

"I want to see it. I want to see what it looks like. I want to understand where it is, and I think this will help."

Brown and planetary scientist Konstantin Batygin feel certain about their prediction, which at first seemed unbelievable to even them.

"For the first time in more than 150 years, there's good evidence that the planetary census of the solar system is incomplete," Batygin said, referring to Neptune's discovery as Planet 8.

Once it's detected, Brown insists there will be no Pluto-style planetary debate. Brown ought to know; he's the so-called Pluto killer who helped lead the charge against Pluto's planetary status in 2006. (Once Planet 9, Pluto is now officially considered a dwarf planet.)

"THIS is what we mean when we say the word 'planet,' " Brown said.

Brown and Batygin believe it's big - 10 times more massive than Earth - and unlike Pluto, dominates its cosmic neighborhood. Pluto is a gravitational slave to Neptune, they pointed out.

Another scientist, Alan Stern, said he's withholding judgment on the planet prediction. He is the principal scientist for NASA's New Horizons spacecraft, which buzzed Pluto last summer in the first-ever visit from Planet Earth. He still sees Pluto as a real planet - not a second-class dwarf.

"This kind of thing comes around every few years. To date, none of those predicts have been borne out by discoveries," Stern said in an email Wednesday. "I'd be very happy if the Brown-Batygin were the exception to the rule, but we'll have to wait and see. Prediction is not discovery."

Brown and Batygin shaped their calculation on the fact that six objects in the icy Kuiper Belt, or Twilight Zone on the far reaches of the solar system, appear to have orbits influenced by only one thing: a real planet. The vast, mysterious Kuiper Belt is home to Pluto as well.

Brown actually discovered one of these six objects more than a decade ago, Sedna, a large minor planet.

"What we have found is a gravitational signature of Planet 9 lurking in the outskirts of the solar system,' Batygin said. The actual discovery, he noted, will be "era-defining."

Added Brown: "We have felt a great disturbance in the force."

Scott Sheppard of the Carnegie Institution for Science in Washington said Brown and Batygin's effort takes his own findings to "the next level." Two years ago, he and a colleague suggested a possible giant planet.

"I find this new work very exciting," Sheppard said in an email. "It makes the distant Super-Earth planet in our solar system much more real. I would say the odds just went from 50 percent to 75 percent that this distant massive planet is real."

Depending on where this Planet 9 is in its egg-shaped orbit, a space telescope may be needed to confirm its presence, the researchers said. Or good backyard telescopes may spot it, they noted, if the planet is relatively closer to us in its swing around the sun. It's an estimated 20 billion to 100 billion miles away.

The Caltech researchers prefer calling it Planet 9, versus the historical term Planet X. The latter smacks of "aliens and the imminent destruction of the Earth," according to Brown.

Who knows, there could even be a Planet 10 out there well beyond No. 9, but there aren't enough data at this point to guess, Brown said.

The last real planet to be discovered in our solar system was Neptune in 1846. Pluto's discovery came in 1930; humanity got to see the small icy world and its main moon Charon up close for the first time last July thanks to New Horizons.

The spacecraft, unfortunately, is in the opposite direction of Planet 9, according to the researchers, and thus unable to help in its detection.

Brown realizes skepticism will exist until the planet is actually observed. History is packed with mistaken planet-seekers, he said, and so "standing up and saying we're right this time makes us almost look crazy - except I'm going to stand up and say we're actually right this time."

He couldn't resist this jab on his @plutokiller Twitter account:

"OK, OK, I am now willing to admit: I DO believe that the solar system has nine planets."

Chapter 4

MISSION TO PLUTO

An Atlas V rocket that carried the New Horizons spacecraft to Pluto lifts off at the Cape Canaveral Air Force Station in Florida for the 9 ½ year journey to Pluto, January 19, 2006. (AP Photo/Terry Renna)

EARLY MISSION PLAN
Pasadena, California, Thursday, February 8, 1993

Maverick NASA scientists and engineers, heeding their boss' call for faster and cheaper space missions, want to send two small space probes to study Pluto, the solar system's last known unexplored planet.

The proposal has yet to receive congressional funding or NASA approval as a full-fledged mission. It has revealed tensions between the agency's tradition of building big, expensive spacecraft and those pressing to simplify and speed up such projects and reduce their cost.

Those designing the Pluto mission are following the "faster, better, cheaper" philosophy advocated by Daniel S. Goldin, administrator of the National Aeronautics and Space Administration.

"It's a very young, aggressive, competitive team - guys who are too dumb and young to know we can't do it," said Rich Terrile, the Pluto mission's chief scientist at NASA's Jet Propulsion Laboratory.

"We saw this space program turned into a fat bureaucracy and we're chomping at the bit to recapture the good old days," he said.

The Pluto mission could cost $ 600 million to $ 1 billion in current dollars, depending on whether the probes are launched by Russian or U.S. rockets, respectively. By comparison, NASA expects to spend more than $ 2 billion to launch the Cassini spaceship in 1997 to investigate Saturn.

The proposed mission to Pluto and its moon, Charon, has found favor with Goldin.

Goldin declined to be interviewed by The Associated Press. And his chief spokesman, Jeff Carr, insisted Goldin hasn't made the Pluto mission a priority, but only gave it "higher visibility" as a model for his "campaign to do things faster, better, cheaper."

Daniel S. Goldin, administrator of U.S. National Aeronautics and Space Administration (NASA), June 3, 1997. (AP Photo/Tsugufumi Matsumoto)

That still upsets some researchers who say Goldin, nevertheless, is putting Pluto ahead of more important projects.

Goldin is a Bush administration appointee and his future under President Clinton is uncertain. The same goes for any projects that get his blessing.

Goldin is trying to put public fascination with space exploration on an equal footing with NASA's scientific pursuits, according to Bruce Murray, a California Institute of Technology professor and former director of NASA's Jet Propulsion Laboratory.

The young Pluto mission planners - scientists and engineers mostly in their 30s - want to use two 362-pound, space probes. Each would measure 4 -feet in diameter and 3 feet high and bear just four instruments each, including a tiny television camera.

The probes would be launched around 1999 to reach Pluto between 2005 and 2007, flying within 6,200 miles of the planet. Each probe would study one side of Pluto.

A Pluto mission the old NASA way would cost at least twice as much, take more than 20 years to execute and would employ one- or two-ton spacecraft laden with scientific instruments.

"Instead of sending another houseboat, we're sending very sophisticated desk-sized spacecraft," said Alan Stern, a planetary scientist at the Southwest Research Institute in San Antonio and chairman of one of two NASA advisory panels that support the proposed mission.

Voicing the other side, John Pike, space policy analyst for the Federation of American Scientists in Washington, said, "NASA has a process for selecting missions on the basis of scientific merit, and Pluto was pretty far down on that list."

Learning more about planets already visited, for instance, holds more interest and import for many researchers, he said.

NASA once developed state-of-the art technology. To reduce the risk of failure, NASA's recent planetary spaceships have used existing technology.

If space probes are to reach distant Pluto, however, they must be lightweight and fast. That means using cutting-edge technology, said Robert Staehle, the project's manager.

Pluto, discovered in 1930 by U.S. astronomer Clyde Tombaugh, is the solar system's ninth and smallest known planet at roughly two-thirds the diameter of Earth's moon.

While NASA has sent robotic explorers to every other known planet, it has yet to explore Pluto, named for the Roman god of the underworld, and its moon Charon, named for the mythical boatman who ferried the dead across the River Styx into Pluto's realm.

Usually, Pluto is the outermost planet, but it's long elliptical orbit sometimes bring it closer to the sun than Neptune, as it has been since 1979. It again becomes the outermost planet in 1999.

"It's the Mount Everest of planetary exploration," Staehle said. "It's usually the farthest away. It's the coldest. It's the hardest to see."

Pluto is "the missing piece of the puzzle which helps us understand the origin of the solar system," Terrile said.

Scientists know Pluto is rocky and icy, unlike the other planets in the outer orbits of the sun - Jupiter, Saturn, Uranus and Neptune - which essentially are balls of gas.

The solar system formed 4.6 billion years ago when a swirling disk of dust, ice and gas clumped to form the planets. Many astronomers believe the known solar system may be surrounded by a belt of hundreds, maybe thousands of never-seen objects they call "ice dwarfs," thought to be between comets and planets in size.

Pluto and Charon could turn out to be large ice dwarfs. Scientists said craters on Pluto and Charon may disclose something about the nature of ice dwarfs, if such objects have hit the planet and its moon.

NASA last launched spacecraft toward an unexplored planet in 1977, when the twin Voyager probes sailed off to Jupiter, Saturn, Uranus and Neptune.

The disquiet over Pluto arose some months after NASA managers told Staehle in January 1992 he could start at least planning the mission.

That May, Staehle got Goldin's ear when the NASA chief was in Los Angeles returning an Oscar. Shuttle astronauts had taken a statuette into space as part of a special Academy Award honor last year for George Lucas, creator of the Star Wars film series.

Soon Goldin was touting the project. And soon criticism ensued.

"I object to the offhand way in which the Pluto mission suddenly was raised in priority when Staehle intercepted Dan Goldin in Hollywood," said Clark Chapman of the Planetary Science Institute in Tucson, Arizona.

But taxpayers are fascinated by missions to unexplored planets, so Congress may find the Pluto mission appealing when asked to start financing the project, probably next year, Staehle said.

NASA told Staehle the cost of two Pluto spacecraft can't exceed $ 400 million. Titan 4-Centaur launch vehicles alone cost $ 400 million. That would boost the mission's price to $ 1.2 billion, not including costs for running the mission.

But Staehle said NASA expects to get a price break on the American rockets, bringing the mission cost to under $ 1 billion. Or it may buy cheaper, slower Russian Proton launchers. They would slow the mission's arrival at Pluto to 2007 to 2010, but drop the mission's entire cost to $ 600 million.

REWRITING TEXTBOOKS
Cape Canaveral, Florida, Thursday, January 12, 2006

It will be the fastest spacecraft ever launched, zooming past the moon in nine hours and reaching Jupiter in just over a year at a speed nearly 100 times that of a jetliner.

Model of actual New Horizons spacecraft, December 19, 2005. (NASA/AP)

Its target is Pluto the solar system's last unexplored planet, 3 billion miles from Earth. And the New Horizons spacecraft, set for liftoff on Tuesday, could reach it within nine years.

Pluto, a tiny, icy misfit of a planet some say it's not a planet at all neither resembles the rocky bodies of Mercury, Venus, Earth and Mars, nor the giant gaseous planets of Jupiter, Saturn, Uranus and Neptune. For years after its discovery 75 years ago, it was considered a planetary oddball.

But in recent years, astronomers have come to realize that Pluto's class of planetary bodies, ice dwarfs, isn't so odd after all. In fact, ice dwarfs are the most populous group in the solar system. Now, scientists have a chance to learn more about them and the origins of the planetary system.

"Just as a Chihuahua is still a dog, these ice dwarfs are still planetary bodies," said Alan Stern of the Southwest Research Institute in Boulder, Colo., the mission's principal investigator. "The misfit becomes the average. The Pluto-like objects are more typical in our solar system than the nearby planets we first knew."

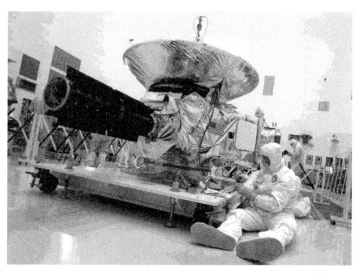

Technicians work on the payload for the New Horizons mission at the Kennedy Space Center in Cape Canaveral, Florida. The mission aboard an Atlas V 551 rocket is set to lift off in January 2006. Travel time to Pluto is from 9 to 14 years, November 4, 2005. (AP Photo/John Raoux)

When the 7-foot-tall New Horizons spacecraft reaches Pluto as early as 2015, the spacecraft will study the ninth planet's large moon, Charon, as well as two other moons just discovered last year. The $700 million mission should provide scientists with a better understanding of the Kuiper Belt, a mysterious region that lies beyond Neptune at the outer limits of the planetary system.

Besides being home to Pluto, the Kuiper Belt is believed to hold thousands of comets and icy planetary objects that make up a third zone of the solar system, the rocky and gaseous planets making up the other two. Scientists believe they can learn about the evolution of the solar system by studying the Kuiper Belt since it possesses debris left over from the formation of the outer solar system. Depending on its fitness after arriving at Pluto, New Horizons will attempt to identify one or two objects in the Kuiper Belt.

"It provides for us a window 4 1/2 billion years back in time to observe the formation conditions of giant planets," Stern said. "This is a little bit about rewriting the textbooks about the outer planets."

A successful journey to Pluto will complete a survey of the planets that NASA began in the early 1960s with the Mariner program's exploration of Mars, Mercury and Venus by unmanned spacecraft. The best images of Pluto currently come from the Hubble Space Telescope, but they suffer from low-resolution fuzziness, making it difficult for scientists to interpret what they're seeing.

The 1,054-pound piano-sized spacecraft will be launched on an Atlas V. The rocket's makers, Lockheed Martin, experienced problems on another Atlas propellant tank similar to the one being flown to Pluto, forcing a delay of New Horizons' launch by several days to give the contractor extra time for inspection.

"Because we have such a long way to go, we put this small spacecraft on one of the largest rockets the U.S. has in its inventory," said project manager Glen Fountain of the Johns Hopkins University Applied Physics Laboratory.

When New Horizons reaches Jupiter in 13 months, it will use that giant planet's gravity as a slingshot, shaving five years off the trip to Pluto. During the trip between Jupiter and Pluto, the probe will go into hibernation, closing down most systems to conserve power. It will send weekly "beeps" back to Earth, providing updates on the vehicle's condition.

An Atlas V rocket that will carry the New Horizons spacecraft on a mission to the planet Pluto lifts off from launch pad 41 at the Cape Canaveral Air Force Station, January 19, 2006. (AP Photo/Terry Renna)

Technicians work on the payload, foreground, for the New Horizons mission at the Kennedy Space Center. In the background is the Payload fairing, which will house the payload, November 4, 2005. (AP Photo/John Raoux)

If the spacecraft is unable to launch during its month long window that closes February 14, the next opportunity is in February

2007, but that would push back an arrival at Pluto to 2020 since New Horizons wouldn't be able to get the gravity assist from Jupiter then.

Powered by nuclear fuel that will produce less energy than is used by two 100-watt lightbulbs, New Horizons is loaded with seven instruments that will be able to photograph the surfaces of Pluto and Charon and examine Pluto's atmospheric composition and structure. Two of the cameras, Alice and Ralph, are named for the bickering couple from television's "The Honeymooners."

Audrey Meadows, left, and Jackie Gleason pose prior to rehearsing for a show aimed at reintroducing 67 segments of "The Honeymooners" TV show at a Miami beach studio, May 1, 1985. (AP Photo/Joe Skipper)

The spacecraft has a thermos-bottle design that will allow it to stay at room temperature. Tucked inside the probe will be a U.S. flag and a CD containing about a half million names of ordinary citizens who signed up on a NASA Web site.

Pluto and the Kuiper Belt have been full of surprises in recent years.

Scientists discovered in 2001 that binary objects pairs like Pluto and Charon litter the Kuiper Belt, and a year later they learned that Pluto's atmosphere undergoes rapid and dramatic global change. Last summer, scientists discovered Pluto's two extra moons.

Scientists expect more unexpected discoveries from the New Horizons mission.

Said Stern, "You can see why we think it's going to be like kids in a candy shop."

3 BILLION MILE TRIP CARRYING 435,000 NAMES
Orlando, Florida, Thursday, January 12, 2006

Cliff Friedman's name may go far in Dallas real estate circles, but that doesn't come close to the 3 billion-mile trip it will begin in a few weeks.

Friedman's name is one of about 435,000 names placed on a compact disc that will be in the New Horizons spacecraft when it launches from Cape Canaveral this week on a mission to Pluto and the outer edges of the solar system.

"My wife thinks it should be put out a little bit further," said Friedman, 48, a Dallas real estate attorney.

The inclusion of names in the spacecraft is part of a public relations campaign to generate interest in the launch to the last unexplored planet in the solar system.

When it reaches Pluto as early as 2015, the spacecraft will study the ninth planet, its large moon, Charon, as well as two other moons only discovered last year. The $700 million mission also should provide scientists with a better understanding of the Kuiper Belt, a mysterious disc-shaped region that lies beyond Neptune at the outer limits of the planetary system.

Other missions have employed a similar P.R. trick.

A small DVD containing more than 616,000 handwritten signatures was in the Cassini orbiter on its mission to Saturn in 1997. Discs with names also were sent aboard the two Mars Rovers and the Deep Impact spacecraft on its successful mission to collide with a comet.

"This is a great way to get people involved," said Michael Buckley, a spokesman for the Johns Hopkins University Applied Physics Lab, which is managing the spacecraft's operations.

Those interested were able to submit their name to the mission's Web site and print out a certificate that read in part, "Thank you for joining the first mission to the last planet!" The submission deadline was in September.

Friedman knew about the mission since he is friends with its principal investigator, Alan Stern, whose brother is Friedman's law partner.

"Knowing about the mission ... I thought it would be a neat thing to have my name go out there, out to the great beyond," he said. "I think it is a great way to get more interest and involve more people in a mission like this."

SET TO LIFT OFF
Cape Canaveral, Florida, Monday, January 16, 2006

An unmanned NASA spacecraft the size of a piano is set to lift off Tuesday on a nine-year journey to Pluto, the last unexplored planet in the solar system.

Scientists hope to learn more about the icy planet and its large moon, Charon, as well as two other, recently discovered moons in orbit around Pluto.

The $700 million New Horizons mission also will study the surrounding Kuiper Belt, the mysterious zone of the solar system that is believed to hold thousands of comets and other icy objects. It could hold clues to how the planets were formed.

"They finally are going! I can't believe it!" said Patricia Tombaugh, 93, widow of Clyde Tombaugh, the Illinois-born astronomer who discovered Pluto in 1930.

Patricia Tombaugh, 93 of Las Cruces, New Mexico, stands next to a model of the New Horizons spacecraft at the Kennedy Space Center, January 15, 2006. (AP Photo/John Raoux)

Patricia Tombaugh, her two children, and the astronomer's younger sister planned to witness the launch of the New Horizons spacecraft at the Cape Canaveral Air Force Station on Tuesday afternoon.

Pluto is the only planet discovered by a U.S. citizen, though some astronomers dispute Pluto's right to be called a planet. It is an oddball icy dwarf unlike the rocky planets of Mercury, Venus, Earth and Mars and the gaseous planets of Jupiter, Saturn, Uranus and Neptune.

NASA has sent unmanned space probes to every planet but Pluto.

"What we know about Pluto today could fit on the back of a postage stamp," said Colleen Hartman, a deputy associate administrator at NASA. "The textbooks will be rewritten after this mission is completed."

Dr. Dave McComas, the principal investigator for the solar wind analyzer (SWAP), uses a model to explain how the instrument will work aboard the New Horizons spacecraft during a press conference at the Kennedy Space Center, January 15, 2006 (AP Photo/John Raoux)

New Horizons will lift off on an Atlas V rocket, which was rolled
to the launch pad Monday, and speed away from Earth at 36,000
mph, the fastest spacecraft ever launched. It will reach Earth's
moon in about nine hours and arrive in 13 months at Jupiter, where
it will use the giant planet's gravity as a slingshot, shaving five years
off the 3-billion-mile trip.

The launch had drawn protests from anti-nuclear activists because the spacecraft will be powered by 24 pounds of plutonium,
which will produce energy from natural radioactive decay.

NASA and the U.S. Department of Energy have put the probability of an early-launch accident that could release plutonium at 1
in 350. The agencies have brought in 16 mobile field teams that can
detect radiation and 33 air samplers and monitors.

"Just as we have ambulances at football games, you don't expect to use them, but we have them there if we need them," NASA
official Randy Scott said.

DELAY
Cape Canaveral, Florida, Wednesday, January 18, 2006

Bad weather forced NASA to scrub its first planned launch of
an unmanned spacecraft on a nine-year voyage to Pluto, the solar
system's last unexplored planet, and strong winds were threatening
another delay Wednesday.

The launch of the New Horizons probe was called off Tuesday
afternoon after winds at the launch pad exceeded the space agency's
38 mph flight restriction.

"The winds picked up sooner than expected," said MIT scientist
Richard Binzel, one of the mission's investigators. "Blame the meteorologists."

NASA has until February 14 to launch New Horizons on its 3-
billion-mile voyage.

A successful journey to Pluto would complete an exploration of
the planets started by NASA in the early 1960s with unmanned missions to observe Mars, Mercury and Venus.

Pluto is an oddball icy dwarf unlike the rocky planets of Mercury, Venus, Earth and Mars and the gaseous planets of Jupiter,
Saturn, Uranus and Neptune.

It also is the brightest body in a zone of the solar system known as the Kuiper Belt, made up of thousands of icy, rocky objects, including tiny planets whose development was stunted by unknown causes.

Scientists believe studying those "planetary embryos" can help them understand how planets were formed.

An Atlas V rocket that will carry the New Horizons spacecraft on a mission to the planet Pluto sits on the launch pad after the liftoff was scrubbed at the Kennedy Space Center. A storm in Maryland knocked out the power to the John Hopkins Applied Physics Laboratory which is managing the operations of the New Horizons spacecraft, January 18, 2006. (AP Photo/John Raoux)

The New Horizons spacecraft will be launched on an Atlas V rocket that will accelerate away from Earth at 36,000 mph, the fastest launch speed on record.

Once launched, the craft was expected to reach Earth's moon in about nine hours and arrive in 13 months at Jupiter, where it will use the giant planet's gravity as a slingshot, shaving five years off the trip.

SECOND DELAY
Cape Canaveral, Florida, Wednesday, January 18, 2006

NASA scrubbed its launch of an unmanned spacecraft on a nine-year voyage to Pluto for the second day Wednesday, but this time weather in Maryland was to blame.

A storm in Laurel, Md., knocked out power at the John Hopkins University Applied Physics Laboratory, which is managing operations of the New Horizons spacecraft.

A decision on whether to try for a Thursday launch depended on whether backup power could be restored at the Maryland facility. That decision would be made late Wednesday, said Alan Stern, the mission's principal investigator.

"I ... was not comfortable with launching without backup power," Stern said. "I've been working on this for 17 years ... Two or three days doesn't mean a hill of beans."

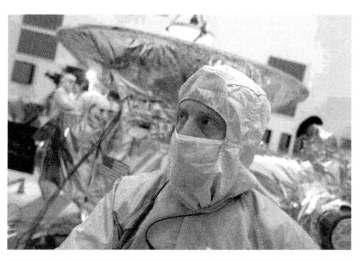

Alan Stern, principal investigator of the New Horizons mission to the planet Pluto, checks progress on assembly of the mission payload at the Kennedy Space Center in Cape Canaveral, November 4, 2005. (AP Photo/John Raoux)

The space agency has until mid-February to launch the piano-sized spacecraft, but a launch in January would allow the spacecraft to use Jupiter's gravity to shave five years off the 3-billion-mile trip.

The launch of the New Horizons probe had been called off Tuesday afternoon when winds at the launch pad in Cape Canaveral exceeded the space agency's 38 mph flight restriction.

"The winds picked up sooner than expected," said MIT scientist Richard Binzel, one of the mission's investigators. "Blame the meteorologists."

A successful journey to Pluto would complete an exploration of the planets that was started by NASA in the early 1960s with unmanned missions to observe Mars, Mercury and Venus.

Pluto is an oddball icy dwarf unlike the rocky planets of Mercury, Venus, Earth and Mars and the gaseous planets of Jupiter, Saturn, Uranus and Neptune.

It also is the brightest body in a zone of the solar system known as the Kuiper Belt, made up of thousands of icy, rocky objects, including tiny planets whose development was stunted by unknown causes.

Scientists believe studying those "planetary embryos" can help them understand how planets were formed.

The planned launch has drawn attention from opponents of nuclear power because the spacecraft is powered by 24 pounds of plutonium, whose natural radioactive decay will generate electricity for the probe's instruments.

NASA and the Department of Energy estimated the probability of a launch accident that could release plutonium at 1 in 350. As a precaution, the agencies brought in 16 mobile field teams that can detect radiation, plus air samplers and monitors.

BLAST OFF
Cape Canaveral, Florida, Thursday, January 19, 2006

A piano-sized spacecraft blasted off Thursday on a 3-billion mile journey to study Pluto, the solar system's last unexplored planet, and examine a mysterious zone of icy objects at the outer edges of the planetary system.

The New Horizons probe lifted off from the Cape Canaveral Air Force Station at 2 p.m., quickly reaching speeds of 36,000 mph, nearly 100 times faster than a jetliner.

"We have ignition and liftoff of NASA's New Horizons spacecraft on a decade long voyage to visit the planet Pluto and then beyond," said NASA commentator Bruce Buckingham.

It was the swiftest spacecraft ever launched and was expected to reach Earth's moon in nine hours and Jupiter in just over a year.

An Atlas V rocket carrying the New Horizons spacecraft on a mission to the planet Pluto, lifts off from launch pad 41 at the Cape Canaveral Air Force Station in Cape Canaveral, Florida, January 19, 2006. (AP Photo/John Raoux)

The distance involved means scientists won't be able to receive data on Pluto until at least July 2015, the earliest the mission is expected to arrive.

After two delays first because of strong winds Tuesday at the launch pad, and then because of a power outage Wednesday at the spacecraft's control center in Maryland New Horizons got off the ground.

"It looked beautiful," said Ralph McNutt Jr. of the Johns Hopkins University of Applied Physics Laboratory, one of the mission scientists. "I was getting a little bit antsy."

Nuclear Engineer Craig Marianno, a contractor for the Department of Energy, demonstrates some of the radiation detection devices that will be used during the New Horizons mission launch. The spacecraft is powered by 24 pounds of plutonium, January 16, 2006. (AP Photo/Terry Renna)

The launch drew attention from opponents of nuclear power because the spacecraft is powered by 24 pounds of plutonium, whose natural radioactive decay will generate electricity for the probe's instruments.

NASA and the Department of Energy had estimated the probability of a launch accident that could release plutonium at 1 in 350. As a precaution, the agencies brought in 16 mobile field teams that can detect radiation and 33 air samplers and monitors.

A successful journey to Pluto would complete an exploration of the planets started by NASA in the early 1960s with unmanned missions to observe Mars, Mercury and Venus.

Pluto is the only planet discovered by a U.S. citizen, though some astronomers dispute Pluto's right to be called a planet. It is an oddball icy dwarf unlike the rocky planets of Mercury, Venus, Earth and Mars and the gaseous planets of Jupiter, Saturn, Uranus and Neptune.

Pluto is the brightest body in a zone of the solar system known as the Kuiper Belt, which is made up of thousands of icy, rocky objects including tiny planets whose development was stunted by unknown causes. Scientists believe studying those "planetary embryos" can help them understand how planets were formed.

The New Horizons spacecraft was launched on an Atlas V rocket, NASA's most powerful launch vehicle. Some NASA safety managers had raised concerns about the rocket's fuel tank since a similar test tank failed a factory pressure evaluation. The decision was made to fly since the flight tank was in pristine condition and had no signs of any defects like the ones found on the test tank.

Chapter 5

THE JOURNEY

New Horizons spacecraft on its mission to Pluto, NASA animation, (Video Screenshot/AP)

FASTEST SPACECRAFT EVER LAUNCHED
Cape Canaveral, Florida, Friday, January 20, 2006

The fastest spacecraft ever launched began the first full day of its 3-billion mile journey to Pluto, where it will study the last unexplored planet and the mysterious icy area that surrounds it.

The New Horizons spacecraft blasted off aboard an Atlas V rocket Thursday afternoon in a spectacular start to the $700 million mission. Despite the speed it can reach 36,000 mph it will take 9 1/2 years to reach Pluto and the frozen, sunless reaches of the solar system.

"It looked beautiful," said Ralph McNutt Jr. of the Johns Hopkins University Applied Physics Laboratory, one of the mission's scientists. "I was getting a little bit antsy."

The 1,054-pound spacecraft was loaded with seven instruments that will photograph the surfaces of Pluto and its large moon, Charon, and analyze Pluto's atmosphere. Two of the cameras, Alice and Ralph, are named for the bickering couple from TV's "The Honeymooners."

New Horizons also contained some of the ashes of Clyde Tombaugh, the astronomer who discovered Pluto in 1930. His widow, Patricia Tombaugh, was in tears as she watched the launch from four miles away.

Patricia Tombaugh, 93, attends a press conference at the Kennedy Space Center in Cape Canaveral, Florida after the launch of an Atlas V rocket that will carry the New Horizons spacecraft on a mission to the planet Pluto, January 19, 2006. (AP Photo/John Raoux)

"I got emotional. I really did. I just got carried away," said Tombaugh, 93, of Las Cruces, New Mexico "It was so beautiful and we've waited so long."

NASA had postponed the liftoff two straight days because of wind gusts at the launch pad and a power outage at the spacecraft's control center in Maryland.

Pluto is the solar system's most distant planet and the brightest body in a zone known as the Kuiper Belt, which is made up of thousands of icy, rocky objects, including tiny planets whose development was stunted for unknown reasons. Scientists believe studying those "planetary embryos" can help them understand how planets were formed.

New Horizons spacecraft on its mission to Pluto, NASA animation (Video Screenshot/AP)

Some astronomers question whether Pluto is technically a planet. Pluto is a celestial oddball an icy dwarf unlike the rocky planets of Mercury, Venus, Earth and Mars and the gaseous planets of Jupiter, Saturn, Uranus and Neptune.

"We're realizing just how much there is to the deep, outer solar system," said Alan Stern, the mission's principal investigator. "I think it's exciting that textbooks have to be rewritten, over and over."

Because it was launched in January, the spacecraft will be able to use Jupiter's gravity as a sling to shave five years off the trip, allowing it to arrive as early as July 2015.

The probe, powered by 24 pounds of plutonium, will not land on Pluto but will photograph it, analyze its atmosphere and send data back across the solar system to Earth.

The launch went off without incident, to the relief of anti-nuclear activists who had feared an accident could scatter lethal radioactive material.

The probe will rely on the natural decay of the plutonium to generate electricity for its instruments. NASA and the Energy Department had put the chances of a launch accident that could release radiation at 1 in 350. As a precaution, the agencies brought in 16 mobile field teams that can detect radiation and 33 air samplers and monitors.

"Certainly there are feelings of relief that we didn't have to actually execute any of our contingency plans," said Bob Lay, emergency management director for surrounding Brevard County.

NASA administrator Michael Griffin said he had an answer for those who may question spending $700 million on the mission to a place in space too far away to observe in any detail from Earth.

NASA Administrator Michael Griffin, left, and Deputy NASA Administrator Shana Dale answer questions at a press conference prior to the scheduled launch of the New Horizons spacecraft at the Kennedy Space Center, January 17, 2006. (AP Photo/Terry Renna)

"Of what value do you think it might be to be able to study the primordial constituents from which the solar system and all the planets and we, ourselves, were formed?" Griffin said.

JUPITER BOOST

Laurel, Maryland, Wednesday, February 28, 2007

NASA's New Horizon's space probe was pointed toward Pluto and the frozen, sunless reaches of the solar system on a nine-year journey after getting a gravity boost Wednesday from Jupiter.

The fastest spacecraft ever launched was within a million and a half miles of Jupiter early Wednesday, giving scientists a close-up look at the giant gaseous planet and its moons.

Mission managers at the Johns Hopkins Applied Physics Laboratory in Laurel, Md. waited for the first signals from the spacecraft after it emerged from behind the giant planet.

Tim Miralles, a flight controller at the Johns Hopkins University Applied Physics Laboratory, who also helped to build and test the New Horizons spacecraft, watches as data comes in from the spacecraft after it passed Jupiter, February 28, 2007. (AP Photo/Jacquelyn Martin)

Missions operations manager Alice Bowman ran through the status checklist, then cheers rang out when she declared: "The spacecraft is outbound from Jupiter and we're on our way to Pluto."

The probe, now accelerating to more than 52,000 mph, was designed and built at the lab in Laurel and tested at NASA's Goddard

Space Flight Center in Greenbelt. It will be making more than 700 separate science observations of the Jupiter system from January through June.

Project manager Glen Fountain said the craft will also travel through Jupiter's 'magneto tail,' which he described as a "tear drop-shaped bubble of plasma," streaming away from the planet into space.

"This is a region never before seen," Fountain said.

New Horizons began its 3 billion mile, 9 1/2-year journey to Pluto in January 2006, where it will study this unexplored dwarf planet and the mysterious icy area that surrounds it. When New Horizons took off, Pluto was a full-fledged planet, but astronomers downgraded it in August. The space probe's arrival there is expected in July 2015.

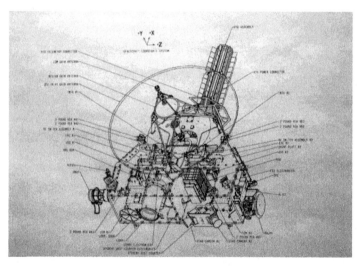

A drawing of the New Horizons spacecraft is tacked to the wall in the Johns Hopkins University Applied Physics Laboratory while the mission operational team receives data from the spacecraft as it passes Jupiter, February 28, 2007. (AP Photo/Jacquelyn Martin)

COSMIC COINCIDENCE
Cape Canaveral, Florida, Monday, August 25, 2014

NASA calls it a cosmic coincidence.

On Monday, NASA's New Horizons spacecraft crossed the orbit of Neptune on its way to Pluto. The celestial milestone occurred on the 25th anniversary of Voyager 2's historic flyby of Neptune.

It's the last major intersection for New Horizons, which is due at Pluto next summer after nearly a decade of travel.

Neptune wasn't exactly close to the spacecraft Monday. In fact, the planet was 2.5 billion miles away.

Scientist Alan Stern of the Southwest Research Institute in Boulder, Colorado, says this will be the first opportunity in a generation to explore a new planetary system up close. New Horizons will study not only mysterious Pluto, but also its moons, some of which might still be hiding from us.

SHOWTIME FOR PLUTO
Cape Canaveral, Florida, Friday, January 23, 2015

It's showtime for Pluto.

NASA's New Horizons spacecraft has traveled 3 billion miles and is nearing the end of its nine-year journey to Pluto. Sunday, it begins photographing the mysterious, unexplored, icy world once deemed a planet.

The first pictures will reveal little more than bright dots - New Horizons is still more than 100 million miles from Pluto. But the images, taken against star fields, will help scientists gauge the remaining distance and keep the baby grand piano-sized robot on track for a July flyby.

It is humanity's first trip to Pluto, and scientists are eager to start exploring.

"New Horizons has been a mission of delayed gratification in many respects, and it's finally happening now," said project scientist Hal Weaver of Johns Hopkins University's Applied Physics Laboratory.

"It's going to be a sprint for the next seven months, basically, to the finish line," he said Friday. "We can't wait to turn Pluto into a real world, instead of just a little pixelated blob."

Launched from Cape Canaveral in January 2006 on a $700 million mission, New Horizons awoke from its last hibernation period early last month. Flight controllers have spent the past several weeks getting the spacecraft ready for the final but most important leg of its journey.

"We have been working on this project, some people, for over a quarter of their careers, to make this mission happen," said project manager Glen Fountain of the Applied Physics Lab, "and now we're about to hit the mother lode."

The spacecraft's long-range reconnaissance imager will take hundreds of pictures of Pluto over the coming months. It snapped pictures last summer, before going into hibernation, but these new ones should be considerably brighter. It will be a few days before the new images are beamed back to Earth; scientists expect to release them publicly in early February.

By May, New Horizons' photos should equal and then surpass the ones taken by the Hubble Space Telescope, with pictures of the plutoid and its moons improving with each passing day.

The real payoff will come when New Horizons flies by Pluto on July 14 at a distance of 7,700 miles and speed of nearly 31,000 mph. It will whip past Charon, Pluto's largest moon, from 18,000 miles out.

Scientists have no idea, really, what Pluto looks like way out in the Kuiper Belt beyond Neptune's orbit, home to little icy objects galore.

Pluto is the biggest object in the Kuiper Belt. Together with mega-moon, Charon, roughly half Pluto's size, the two orbs could fit inside the United States with room to spare. Five moons have been found so far around Pluto. More could be lurking out there, awaiting discovery by New Horizons.

The Applied Physics Lab in Laurel, Maryland, designed and built New Horizons, and is now managing the mission for NASA.

Pluto was still officially a planet, No. 9 in the solar system lineup, when New Horizons departed Earth. It was the only planet in our solar system yet to be explored. But seven months later, the International Astronomical Union stripped Pluto of its planethood, classifying it instead as a dwarf planet. Later came the term, plutoid.

Some scientists are hoping Pluto's upcoming close-up - and expected cosmic buzz - may prompt the group to reverse its decision.

The nature of science, after all, is fluid, as even the astronomical union maintains.

Streator, Illinois, - hometown of Pluto's discoverer, the late astronomer Clyde Tombaugh - already has declared 2015 the "Year of Pluto." Tombaugh spotted Pluto in 1930.

New Horizons may, indeed, "turn the tide in some people's opinions into the other camp," Weaver said. "But that's not really so important."

More important, he said, is finding out "what does Pluto really look like."

AT PLUTO'S DOORSTEP
Cape Canaveral, Florida, Tuesday, June 16, 2015

NASA's New Horizons spacecraft is at Pluto's doorstep, following an incredible journey of nine years and 3 billion miles.

Pluto Mission spacecraft, December 20, 2000. (NASA Artist Conception/AP)

Four weeks from Tuesday - on July 14 - New Horizons will make its closest approach to Pluto. The spacecraft will fly within

7,750 miles, inside the orbits of Pluto's five known moons. That's the approximate distance between Seattle and Sydney.

It will be the first spacecraft to explore the tiny, icy world once considered a full-fledged planet.

As of Tuesday, New Horizons was just over 20 million miles from Pluto. That's closer than Earth is to neighbor Venus, at their closest point. Flight controllers fired a thruster on the spacecraft over the weekend to fine-tune its path.

"This is one charged-up team," principal investigator Alan Stern of the Southwest Research Institute in Boulder, Colorado said last week. "They know that they're getting to do something very special because nothing like this has happened" since Voyager 2's flyby of Neptune in 1989. NASA's first interplanetary success was at Venus, with Mariner 2 in 1962.

Stern added: "We're going to turn a point of light into a planet and its moons overnight in the next month."

The Johns Hopkins University's Applied Physics Laboratory in Laurel, Maryland, is operating the spacecraft for NASA. The lab also designed and built the relatively lightweight craft, about the size of a baby grand piano. It carries seven science instruments; the cameras have been photographing the planet since January.

The latest pictures, taken at the end of May and beginning of June, show large dark regions toward the bottom of Pluto. Scientists are eager to learn the size and shape of these dark spots, as well as their exact location. Images will keep improving with every step closer to Pluto.

"It's very fascinating to see this level of detail," deputy project scientist Cathy Olkin said during an update broadcast Tuesday.

Pluto was discovered by the late American astronomer Clyde Tombaugh in 1930. Its moons, the fifth unmasked as recently as 2012, also bear names related to the mythological underworld: big moon Charon and mini-moons Nix, Hydra, Kerberos and Styx.

More moons could be out there.

"We're going to write the textbook. We know very little about the Pluto system now," Stern said. "It's really a mission of raw exploration, flying into the unknown to see what's there."

New Horizons' $700 million mission began with a 2006 launch from Cape Canaveral, Florida.

'SPEED BUMP'
Cape Canaveral, Florida, Tuesday, July 6, 2015

NASA's New Horizons spacecraft is on track to sweep past Pluto next week despite hitting a "speed bump" that temporarily halted science collection.

A computer overload prompted the spacecraft to partially shut down on July 4th - just days before the first-ever close flyby of Pluto. Flight controllers managed to regain contact with the spacecraft in just over an hour and correct the tense situation, occurring after a relatively quiet journey of 3 billion miles and 9½ years.

Pluto Mission satellite, February 4, 1993. (NASA Artist Conception/AP)

"We're on to Pluto!" NASA's director of planetary science, Jim Green, assured journalists Monday.

New Horizons will pass closest to Pluto on July 14. The spacecraft will come within 7,750 miles of the mysterious, tiny, icy world out on the fringes of the solar system.

About 2½ days of science observations were lost because of the problem. That represents about 30 observations out of 500 planned over the next week.

Principal scientist Alan Stern said it was more important to re-cover the spacecraft than worry about some lost observations of a Pluto still several million miles away. Data collection is expected to resume Tuesday, one week before the long-awaited flyby.

"While we prefer that this event hadn't occurred ... this is a speed bump in terms of the total return that we expect from this flyby," said Stern, who is with the Southwest Research Institute in Boulder, Colorado. Pluto and its big moon Charon already are surprising scientists with their surface appearances, "and we're excited to get back to that," he added Monday.

Stern as well as others involved in the mission said they do not expect the problem to re-occur. The main computer was multi-tasking in preparation for the big event coming up - and dealing with heavier, more complex data loads than expected - when the trouble arose. The spacecraft went into its so-called safe mode, and science operations ceased. Contact was restored through its backup computer.

No changes were made to the flyby plan as a direct result of Saturday's problem, according to officials.

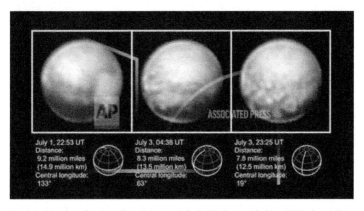

This combination of images from July 1 to July 3, 2015, provided by NASA shows Pluto at different distances from the New Horizons spacecraft. NASA's New Horizons spacecraft is on track to sweep past Pluto next week despite hitting a "speed bump" that temporarily halted science collection, July 6, 2015. (NASA via AP)

New Horizons - about the size of a baby grand piano - was launched from Cape Canaveral, Florida, in 2006. It was designed and built by Johns Hopkins University's Applied Physics Laboratory in Laurel, Maryland, which also is managing the $700 million mission for NASA. That's where the New Horizons flight control team is based.

TOMBAUGH TRAVELS TO PLUTO
Cape Canaveral, Florida, Sunday, July 12, 2015

Come Tuesday, Clyde Tombaugh will pass within 7,800 miles of the icy world he discovered 85 years ago.

His ashes are flying on NASA's New Horizons spacecraft on humanity's first journey to Pluto.

New Horizons also is carrying a 1991 U.S. postage stamp that's about to become obsolete - it trumpets "Pluto Not Yet Explored" - as well as two state quarters, one representing Florida, home of the launch site, and the other Maryland, headquarters for the spacecraft developers and flight control.

In all, nine small mementos are tucked aboard New Horizons. There's a good reason there are nine.

When New Horizons rocketed away from Cape Canaveral on January 19, 2006, Pluto was the ninth planet in our solar system. It was demoted to dwarf planet a scant seven months later.

Tombaugh's widow and two children offered up an ounce of his ashes for the journey to Pluto. The ashes of the farm boy-turned-astronomer are in a 2-inch aluminum capsule inscribed with these words:

"Interned herein are remains of American Clyde W. Tombaugh, discoverer of Pluto and the solar system's 'third zone.' Adelle and Muron's boy, Patricia's husband, Annette and Alden's father, astronomer, teacher, punster, and friend: Clyde Tombaugh (1906-1997)"

Annette Tombaugh-Sitze and her younger brother Alden, now in their 70s, plan to be at the flight operation base at Johns Hopkins University's Applied Physics Laboratory in Laurel, Maryland, for Tuesday's historic encounter. Their mother died in 2012 at age 99.

"I think my dad would be thrilled with the New Horizons. I mean, who wouldn't be?" Annette says in a NASA interview posted online. "When he looked at Pluto, it was just a speck of light."

An aluminum canister containing the ashes of Clyde Tombaugh is attached to NASA's New Horizons spacecraft, (Johns Hopkins University Applied Physics Laboratory via AP)

As for the 29-cent stowaway stamp, Pluto is depicted as grayish with orange flecks, an artist's rendering based on what NASA knew about the tiny orb prior to 1991, which wasn't much.

New Horizons' better and better views reveal a copper-colored, icy bright world.

"No stamp has ever traveled this far!" Mark Saunders, a spokesman for the U.S. Postal Service, said in an email last week.

A small cutout of SpaceShipOne is attached to New Horizons; the first manned private space plane achieved suborbital flights in 2004 and won the $10 million Ansari X Prize.

Annette and Alden Tombaugh, the children of American astronomer Clyde Tombaugh, talk at the New Horizons Pluto flyby event at the Johns Hopkins University Applied Physics Laboratory. The United States is now the only nation to visit every single planet in the solar system, July 14, 2015. (Bill Ingalls/NASA via AP)

Also on the spacecraft are two U.S. flags as well as two CDs. One contains the photos of team members.

The other contains 434,738 names of people who signed up online in advance, including this reporter, holder of Certificate No. 64,646.

PREPARE TO BE AMAZED
Cape Canaveral, Florida, Monday, July 13, 2015

Pluto, reveal thyself, and Earthlings, enjoy the show.

On Tuesday, NASA's New Horizons spacecraft will sweep past Pluto and present the previously unexplored world in all its icy glory.

It promises to be the biggest planetary unveiling in a quarter-century. The curtain hasn't been pulled back like this since NASA's Voyager 2 shed light on Neptune in 1989.

Now it's little Pluto's turn to shine way out on the frigid fringes of our solar system.

New Horizons has traveled 3 billion miles over 9½ years to get to this historic point. The fastest spacecraft ever launched, it carries the most powerful suite of science instruments ever sent on a scouting and reconnaissance mission of a new, unfamiliar world.

Guarantees principal scientist Alan Stern, "We're going to knock your socks off."

The size of a baby grand piano, the spacecraft will come closest to Pluto on Tuesday morning - at 7:49 a.m. EDT. That's when New Horizons is predicted to pass within 7,767 miles of Pluto. Fourteen minutes later, the spacecraft will zoom within 17,931 miles of Charon, Pluto's jumbo moon.

For the plutophiles among us, it will be cause to celebrate, especially for those gathered at the operations center at Johns Hopkins University's Applied Physics Laboratory in Laurel, Maryland. The lab designed and built the spacecraft for NASA, and has been managing its roundabout route through the solar system.

"What NASA's doing with New Horizons is unprecedented in our time and probably something close to the last train to Clarksville, the last picture show, for a very, very long time," says Stern, a planetary scientist with the Southwest Research Institute in Boulder, Colorado.

It is the last stop in NASA's quest to explore every planet in our solar system, starting with Venus in 1962. And in a cosmic coincidence, the Pluto visit falls on the 50th anniversary of the first-ever flyby of Mars, by Mariner 4.

Yes, we all know Pluto is no longer an official planet, merely a dwarf, but it still enjoyed full planet status when New Horizons rocketed from Cape Canaveral, Florida, on January 19, 2006. Pluto's demotion came just seven months later, a sore subject still for many.

"We're kind of running the anchor leg with Pluto to finish the relay," Stern says.

The sneak peeks of Pluto in recent weeks are getting "juicier and juicier," says Johns Hopkins project scientist Hal Weaver. "The science team is just drooling over these pictures."

The Hubble Space Telescope previously captured the best pictures of Pluto. If the pixelated blobs of pictures had been of Earth, though, not even the continents would have been visible.

The New Horizons team is turning "a point of light into a planet," Stern says.

An image released last week shows a copper-colored Pluto bearing, a large, bright spot in the shape of a heart.

New Horizons Scientist Hal Weaver, left, talks with Mark Holdridge, center, and Mike Buckley in Laurel, Maryland as they await information from the New Horizons spacecraft as it passes Pluto, July 14, 2015. (AP Photo/Gail Burton)

Scientists expect image resolution to improve dramatically by Tuesday. The 7,767-mile span at closest approach is about the distance between Seattle and Sydney.

New Horizons, weighing less than 1,000 pounds including fuel, has seven instruments that will be going full force during the encounter. It's expected to collect 5,000 times as much data, for instance, as Mariner 4.

"We're going to rewrite the book," Weaver says. "This is it - this is our once-in-a-lifetime opportunity to see it."

The team gets one crack at this.

"We're trying to hit a very small box, relatively speaking," says Mark Holdridge, the encounter mission manager. "It's 60 by 90 miles, and we're going 30,000 mph, and we're trying to hit that box within a plus or minus 100 seconds."

The only planet in our solar system discovered by an American, Pluto actually is a mini solar system unto itself. Pluto - just two-thirds the size of our own moon - has big moon Charon that's just over half its size, as well as baby moons Styx, Nix, Hydra and Kerberos. The names are associated with the underworld in which the mythological god, Pluto, reigned. New Horizons will observe each known moon and keep a lookout for more.

Scientists involved in the $700 million effort want to get a good look at Pluto and Charon, and get a handle on their surfaces and chemical composition. They also plan to measure the temperature and pressure in Pluto's nitrogen-rich atmosphere and determine how much gas is escaping into space. Temperatures can plunge to nearly minus-400 degrees.

Bill McKinnon, a New Horizons team member from Washington University in St. Louis, Missouri, expects to see craters and possible volcanic remnants. A liquid ocean and a rocky core may lie beneath the icy shell.

"Anybody who thinks that when we go to Pluto, we're going to find cold, dead ice balls is in for a rude shock," McKinnon says. "I'm really hoping to see a very active and dynamic world."

Pluto has tantalized astronomers since its 1930 discovery by Clyde Tombaugh using the Lowell Observatory in Flagstaff, Arizona. Some of Tombaugh's ashes are aboard New Horizons. His two children, now in their 70s, plan to be at Johns Hopkins for the encounter.

With its tilted, elongated 248-year orbit, Pluto has made it only a third of the way around the sun since its discovery. The amount of sunlight that reaches Pluto is so dim that at high noon it looks like twilight here on Earth. The massive surrounding Kuiper Belt, in fact, is called the Twilight Zone. The New Horizons team has its eyes on a few much smaller objects in the Kuiper Belt, and is hoping for a mission extension as the spacecraft continues toward the solar system exit on the heels of NASA's Voyagers 1 and 2 and Pioneers 10 and 11.

Members of the New Horizons science team react to seeing the spacecraft's last and sharpest image of Pluto before closest approach at the Johns Hopkins University Applied Physics Laboratory (APL) in Laurel, Maryland, July 14, 2015. (Bill Ingalls/NASA via AP)

For now, signals take 4½ hours to travel one-way between New Horizons and flight controllers in Maryland.

New Horizons' science instruments will be cranked up to collect maximum data Tuesday, leaving no time to send back data. In fact, scientists won't be absolutely certain of success until Tuesday night, 13 hours following New Horizons' closest approach, when it "phones home."

It will be Wednesday before the closest of Pluto's close-ups are available for release. And it will be well into next year - October 2016 - before all the anticipated data are transmitted to Earth.

"We're all going to have to be patient," urges deputy project scientist Cathy Olkin.

For everyone involved, this is a mission of delayed gratification.

BIGGER THAN IMAGINED
Cape Canaveral, Florida, Tuesday, July 14, 2015

Little Pluto is a little bigger than anyone imagined.

On the eve of NASA's historic flyby of Pluto, scientists announced Monday the New Horizons spacecraft has nailed the size of the faraway icy world.

Measurements by the spacecraft set to sweep past Pluto on Tuesday indicate the diameter of the dwarf planet is 1,473 miles, plus or minus 12 miles. That's about 50 miles bigger than previous estimates in the low range.

Principal scientist Alan Stern said this means Pluto has a lower density than thought, which could mean an icier and less rocky interior.

New Horizons' 3 billion-mile, 9½-year journey from Cape Canaveral, Florida, culminates Tuesday morning when the spacecraft zooms within 7,767 miles of Pluto at 31,000 mph.

Mission managers said there's only one chance in 10,000 something could go wrong, like a debilitating debris strike, this late in the game. But Stern cautioned: "We're flying into the unknown. This is the risk we take with all kinds of exploration."

"It sounds like science fiction, but it's not," Stern said as he opened a news conference at mission headquarters in Maryland. "Tomorrow morning a United States spacecraft will fly by the Pluto system and make history."

Discovered in 1930, Pluto is the last planet in our solar system to be explored. It was a full-fledged planet when New Horizons rocketed away in 2006, only to become demoted to dwarf status later that year.

New Horizons has already beamed back the best-ever images of Pluto and big moon Charon on the far fringes of the solar system.

"The Pluto system is enchanting in its strangeness, its alien beauty," said Stern, a planetary scientist at Southwest Research Institute in Boulder, Colorado.

With the encounter finally at hand, it all seems surreal for the New Horizons team gathered at Johns Hopkins University's Applied Physics Laboratory. The energy there Monday was described as electric.

Project manager Glen Fountain said New Horizons, at long last, is like a freight train barreling down the track, "and you're seeing this light coming at you and you know it's not going to stop, you can't slow it down."

"Of course, the light is Pluto, and we're all excited," Fountain said.

Three new discoveries were revealed Monday, a tantalizing sneak preview as the countdown to closest approach reached the 21-hour mark.

Besides the revised size of Pluto - still a solar system runt, not even one-fifth the size of Earth - scientists have confirmed that Pluto's north pole is indeed icy as had been suspected. It's packed with methane and nitrogen ice.

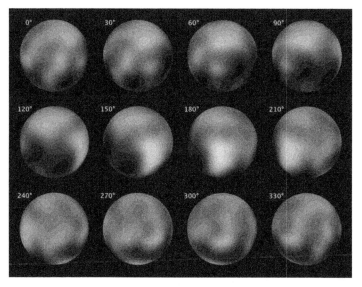

This combination of images made by NASA's Hubble Space Telescope shows Pluto at different angles, 2002 and 2003. (AP Photo/NASA, ESA, M. Buie)

And traces of Pluto's nitrogen-rich atmosphere have been found farther from the dwarf planet than anticipated. New Horizons detected lost nitrogen nearly a week ago.

As for pictures, the resolution is going to increase dramatically. Until New Horizons, the best pictures of Pluto came from the Hubble Space Telescope. Hubble did its best from Earth orbit, but managed to produce only crude pixelated blobs of the minuscule world.

The New Horizons spacecraft is the size of a baby grand piano with a salad bowl - the dish antenna - on top. It will come closest to Pluto at 7:49 a.m. EDT Tuesday. Thirteen hours later, around 9 p.m. EDT, flight controllers will learn if everything went well. The spacecraft will have sent the confirmation signal 4½ hours earlier; that's the one-way, speed-of-light, data-transit time between New Horizons and Earth.

Stern expects "a little bit of drama" during closest approach, when the spacecraft is out of touch with ground controllers. New Horizons cannot make observations and send back data at the same time, so scientists opted for maximum science during those most critical hours.

Pluto is the largest object in the so-called Kuiper Belt, considered the third zone of the solar system after the inner rocky planets and outer gaseous ones. This unknown territory is a shooting gallery of comets and other small bodies; every time one of these wayward objects smack one of Pluto's five known moons, the ejected material ends up in orbit around Pluto, thus the debris concern. An extension of the $720 million mission, not yet approved, could have New Horizons flying past another much smaller Kuiper Belt object, before departing the solar system.

Stern expects two opportunities to celebrate Tuesday: one at the time of closest approach and the other once confirmation is received that New Horizons stayed safe and accomplished its long-awaited mission.

"There is that element of exploration and there's that small element of danger," he said, "so I think we're all going to breathe the final sigh of relief at 9 p.m. and that's when we can really call it a successful flyby."

Chapter 6

PAYOFF

Image of Pluto taken by New Horizons spacecraft, July 24, 2015. (NASA Graphic/AP)

SPOTLIGHT ON PLUTO
Cape Canaveral, Florida, Tuesday, July 14, 2015

The spotlight is bright enough to thaw even Pluto.

Well, not quite, but the tiny, icy world is getting front-page, prime-time attention as NASA's New Horizons spacecraft zooms closer.

Tuesday morning's 31,000-mph flyby - with closest approach at 7,767 miles - is expected to open up new ground on the last un-explored planetary territory of our solar system.

"Turning little dots, little points of light into planets," is what New Horizons, on the road for 9½ years and 3 billion miles, is all about, principal scientist Alan Stern said Monday.

Here's a rundown on Pluto, a 20th-century discovery about to become the 21st-century darling of astronomers:

Discovery

Pluto is the only planet (OK, now former planet) in our solar system discovered by an American. Astronomer Clyde Tombaugh spotted the dot in 1930 from Lowell Observatory in Flagstaff, Arizona. The name Pluto came from a British schoolgirl, Venetia Burney, then 11, based on the mythological god of the underworld.

Tombaugh died at age 90 in 1997, nine years before New Horizons took flight. A smidgen of his ashes is on board. Burney died in 2009, also at age 90. A student-built dust counter aboard New Horizons - from the University of Colorado at Boulder - is named after her.

Five Moons

Big moon Charon was discovered in 1978 by Americans using the U.S. Naval Observatory in Flagstaff, followed by little moons Nix and Hydra in 2005, Kerberos in 2011 and Styx in 2012. The Hubble Space Telescope revealed all four baby moons. Astronomers stuck to underworld undertones when it came to the names.

New Horizons will hunt for more moons, but at this point, they would have to be pretty elusive - scientists guess probably less than a mile across. The Pluto empire, complete with six bodies, at least for now, is like its own mini solar system.

Forget The Sunglasses

Pluto is so far from the sun - between 2.8 billion and 4.6 billion miles - that twilight reigns. At high noon on Pluto, it looks as though it would be dawn or dusk on Earth. And let's not forget the frigid weather, given its distance from the sun. Temperatures can plunge to minus-400 degrees.

Pluto's orbit is extremely oblong, plus it's tilted. It takes 248 years for Pluto to orbit the sun. Thus, it's only made it about one-third of the way around the sun since its discovery in 1930. Every so often, Neptune's orbit exceeds Pluto's, putting Neptune slightly farther out.

First A Planet, Then It's Not

Pluto is the only planet to get kicked out of the solar system club.

Just seven months after New Horizons rocketed away from Cape Canaveral, Florida, the International Astronomical Union declassified Pluto as the ninth planet for technical reasons. Instead, it became a dwarf planet. The decision left the solar system with eight full-fledged planets, with Mercury replacing Pluto as the smallest.

On Monday, scientists said measurements by New Horizons showed Pluto to be 1,473 miles in diameter, a little bigger than earlier estimates.

The Twilight Zone

Pluto is the biggest object in the icy Kuiper Belt, also known as the third zone after the inner rocky planets and outer gaseous giants. It's also called the Twilight Zone because of its great distance from the sun.

The Kuiper Belt (pronounced KIE-per) is full of comets and other small frosty objects. It's named after the late Dutch-American astronomer Gerard Kuiper, who proposed a bevy of small bodies beyond Neptune back in the 1950s. The New Horizons team hopes to go after a smaller Kuiper Belt object following the Pluto flyby, provided a mission extension is approved.

JUBILATION
Cape Canaveral, Florida, Wednesday, July 15, 2015

NASA's New Horizons spacecraft got humanity's first up-close look at Pluto on Tuesday, sending word of its triumph across 3 billion miles to scientists waiting breathlessly back home.

Confirmation of mission success came 13 hours after the actual flyby and, after a day of both jubilation and tension, allowed the New Horizons team to finally celebrate in full force.

"This is a tremendous moment in human history," John Grunsfeld, NASA's science mission chief, said at a news conference.

Principal scientist Alan Stern asked the entire New Horizons team in the audience to stand: "We did it! Take a bow!"

New Horizons team members and guests watch a live feed of the Mission Operations Center (MOC) at the Johns Hopkins University APL as the team waits to receive confirmation from the spacecraft that it has completed the flyby of Pluto, July 14, 2015. (Joel Kowsky/NASA via AP)

The unprecedented encounter was the last stop on NASA's grand tour of our solar-system's planets over the past half-century. The journey began 9½ years ago, back when Pluto was still considered a full-fledged planet.

Tuesday morning, a cheering, flag-waving celebration swept over the mission operations center in Maryland at the time of closest approach. But until New Horizons phoned home Tuesday night, there was no guarantee the spacecraft had buzzed the small, icy, faraway - but no longer unknown - world.

NASA said the spacecraft - the size of a baby grand piano - swept to within 7,700 miles of Pluto at 31,000 mph. It was programmed to then go past the dwarf planet and begin studying its far side.

To commemorate the moment of closest approach, scientists released the best picture yet of Pluto, taken on the eve of the flyby.

Even better images will start "raining" down on Earth beginning Wednesday, promised principal scientist Alan Stern. But he had cautioned everyone to "stay tuned" until New Horizons contacted home.

Jim Green, NASA Planetary Science Division director, center, and other New Horizons team members count down to the spacecraft's closest approach to Pluto at the Johns Hopkins University Applied Physics Laboratory (APL) in Laurel, Maryland. The moment of closest approach for the New Horizons spacecraft came around 7:49 a.m. EDT Tuesday, culminating an epic journey from planet Earth that spanned an incredible 3 billion miles and 9½ years, July 14, 2015. (Bill Ingalls/NASA via AP)

It takes 4½ hours for signals to travel one-way between New Horizons and Earth. The message went out late in the afternoon during a brief break in the spacecraft's data-gathering frenzy. The New Horizons team kept up a confirmation countdown, noting via Twitter when the signal should have passed the halfway point, then Jupiter's orbit.

The uncertainty added to the drama. "This is true exploration," cautioned Stern, a Southwest Research Institute planetary scientist.

Among the possible dangers: cosmic debris that could destroy the mission. But with the chances of a problem considered extremely low, scientists and hundreds of others assembled at Johns Hopkins University's Applied Physics Laboratory erupted in jubilation when the moment of closest approach occurred at 7:49 a.m. EDT. The lab is the spacecraft's developer and manager.

The scene repeated itself a little before 9 p.m. EDT.

This time, the flight control room was packed compared with earlier, when it was empty because New Horizons was out of touch and operating on autopilot.

"We have a healthy spacecraft," announced mission operations director Alice Bowman. She was drowned out by cheers and applause; Stern ran over to give her a hug.

Later, Grunsfeld told reporters, "The spacecraft is full of images. We can't wait. We've opened up a new realm of the solar system."

Added NASA Administrator Charles Bolden: "What a phenomenal day."

Carey Lisse, an instrument scientist on the New Horizons science team reacts after the team received confirmation from the spacecraft that it has completed the flyby of Pluto, July 14, 2015. (Joel Kowsky/NASA via AP)

Joining in the daylong hoopla were the two children of the American astronomer who discovered Pluto in 1930, Clyde Tombaugh. (Some of his ashes are aboard the spacecraft.) Other celestial-minded VIPs included James Christy, discoverer of Pluto's big moon Charon, and Sylvia Kuiper des Tombe, daughter of Dutch-American Gerard Kuiper for whom the mysterious zone surrounding Pluto is named. Some Pluto Children - born January 19, 2006, the very day New Horizons departed Earth - also were in the audience.

Throughout the day - coincidentally the 50th anniversary of the first close-up pictures of Mars from Mariner 4 - the White House and Congress offered congratulations, and physicist Stephen Hawking was among the scientists weighing in. President Barack

Obama sent his best Tuesday night with a tweet: "Pluto just had its first visitor!"

"Hey, people of the world! Are you paying attention?" planetary scientist Carolyn Porco, part of the New Horizons' imaging team, said on Twitter. "We have reached Pluto. We are exploring the hinterlands of the solar system. Rejoice!"

Chaim Loeung, 1, gets Pluto painted on his face at Lowell Observatory in Flagstaff, Arizona, July 14, 2015. (AP Photo/Felicia Fonseca)

The U.S. is now the only nation to visit every planet in the solar system. Pluto was No. 9 in the lineup when New Horizons left Cape Canaveral, Florida, but was demoted seven months later to dwarf status.

Scientists in charge of the $720 million mission hope the new observations will restore Pluto's honor.

Stern and other so-called plutophiles posed for the cameras giving nine-fingers-up "Pluto Salute." And in a nod to that other Pluto, a team member carried a yellow stuffed dog on her shoulder Tuesday night.

The picture of Pluto taken Monday showed a frozen, pockmarked world, peach-colored with a heart-shaped bright spot and darker areas around the equator. It drew oohs and aahs.

"To see Pluto be revealed just before our eyes, it's just fantastic," said Bowman.

The Hubble Space Telescope had offered up the best pre-New Horizons pictures of Pluto, but they were essentially pixelated blobs of light.

Flight controllers held off on having New Horizons send back flyby photos until well after the maneuver was complete; they wanted the seven science instruments to take full advantage of the encounter. After turning toward Earth to send down a snippet of engineering data acknowledging everything was fine, the spacecraft was going to get right back to science work.

New Horizons is also expected to beam back photos of Pluto's big moon, Charon, and observe its four little moons. It will take until late 2016 for all the data to reach Earth.

New Horizons already has confirmed that Pluto is, indeed, the King of the Kuiper Belt. New measurements it made show that Pluto is 1,473 miles in diameter, or about 50 miles bigger than estimated.

Graphic comparing Sedna, Pluto, the Moon and Earth (NASA/AP)

That's still puny by solar-system standards. Pluto is just two-thirds the size of Earth's moon. But it is big enough to be the largest object in the Kuiper Belt, a zone rife with comets and tens of thousands of other small bodies.

Stern and his colleagues wasted no time pressing the U.S. Postal Service for a new stamp of Pluto.

The last one, issued in 1991, consisted of an artist's rendering of the faraway world and the words: "Pluto Not Yet Explored." The words "not yet" were crossed out in a poster held high Tuesday for the cameras.

New Horizons Principal Investigator Alan Stern of Southwest Research Institute (SwRI), Boulder, Colorado, left, Johns Hopkins University Applied Physics Laboratory (APL) Director Ralph Semmel, center, and New Horizons Co-Investigator Will Grundy of the Lowell Observatory hold a print of a U.S. stamp with their suggested update since the New Horizons spacecraft made its closest approach to Pluto, at the Johns Hopkins University Applied Physics Laboratory (APL) in Laurel, Maryland. At center right under the stamp is Annette Tombaugh, daughter of Pluto's discoverer, Clyde Tombaugh, July 14, 2015. (Bill Ingalls/NASA via AP)

'BLOWING MY MIND'
Cape Canaveral, Florida, Wednesday, July 15, 2015

Mankind's first close-up look at Pluto did not disappoint Wednesday: The pictures showed ice mountains on Pluto about as high as the Rockies and chasms on its big moon Charon that appear six times deeper than the Grand Canyon.

Especially astonishing to scientists was the total absence of impact craters in a zoom-in shot of one otherwise rugged slice of Pluto.

That suggests that Pluto is not the dead ice ball many people think, but is instead geologically active even now, its surface sculpted not by collisions with cosmic debris but by its internal heat, the scientific team reported.

Breathtaking in their clarity, the long-awaited images were unveiled in Laurel, Maryland, home to mission operations for NASA's New Horizons, the unmanned spacecraft that paid a history-making flyby visit to the dwarf planet on Tuesday after a journey of 9½ years and 3 billion miles.

"I don't think any one of us could have imagined that it was this good of a toy store," principal scientist Alan Stern said at a news conference. He marveled: "I think the whole system is amazing. ... The Pluto system IS something wonderful."

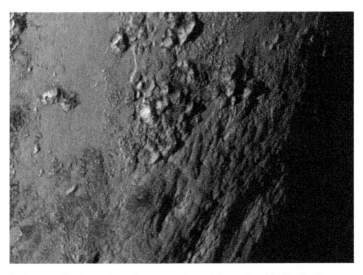

Region near Pluto's equator with a range of mountains captured by the New Horizons spacecraft, July 14, 2015. (NASA/JHUAPL/SwRI via AP)

As a tribute to Pluto's discoverer, Stern and his team named the bright heart-shaped area on the surface of Pluto the Tombaugh Reggio. American astronomer Clyde Tombaugh spied the frozen, faraway world on the edge of the solar system in 1930.

Thanks to New Horizons, scientists now know Pluto is a bit bigger than thought, with a diameter of 1,473 miles, but still just two-thirds the size of Earth's moon. And it is most certainly not frozen in time.

The zoom-in of Pluto, showing an approximately 150-mile swath of the dwarf planet, reveals a mountain range about 11,000 feet high and tens of miles wide. Scientists said the peaks - seemingly pushed up from Pluto's subterranean bed of ice - appeared to be a mere 100 million years old. Pluto itself is 4.5 billion years old.

"Who would have supposed that there were ice mountains?" project scientist Hal Weaver said. "It's just blowing my mind."

John Spencer, like Stern a scientist at the Southwest Research Institute, called it "just astonishing" that the first close-up picture of Pluto didn't have a single impact crater. Stern said the findings suggesting a geologically active interior are going to "send a lot of geophysicists back to the drawing boards."

"It could be a game-changer" in how scientists look at other frozen worlds in the Kuiper Belt on the fringes of our solar system, Spencer said. Charon, too, has a surprisingly youthful look and could be undergoing geologic activity.

"We've tended to think of these midsize worlds ... as probably candy-coated lumps of ice," Spencer said. "This means they could be equally diverse and be equally amazing if we ever get a spacecraft out there to see them close up."

The heat that appears to be shaping Pluto may be coming from the decay of radioactive material normally found in planetary bodies, the scientists said. Or it could be coming from energy released by the gradual freezing of an underground ocean.

As for Charon, which is about half the size of Pluto, its canyons look to be 3 miles to 6 miles deep and are part of a cluster of troughs and cliffs stretching 600 miles, or about twice the length of the Grand Canyon, scientists said.

The Charon photo was taken Monday. The Pluto picture was shot just 1½ hours before the spacecraft's moment of closest approach. New Horizons swept to within 7,700 miles of Pluto during its flyby. It is now 1 million miles beyond it.

Up until this week, the best pictures of Pluto were taken by the Hubble Space Telescope, and they were blurry, pixelated images.

Scientists promised even better pictures for the next news briefing on Friday. Johns Hopkins University's Applied Physics Laboratory is in charge of the $720 million mission.

'BEAUTIFUL EYE CANDY'
Cape Canaveral, Florida, Friday, July 17, 2015

Vast frozen plains exist next door to Pluto's big, rugged mountains sculpted of ice, scientists said Friday, three days after humanity's first-ever flyby of the dwarf planet.

The New Horizons spacecraft team revealed close-up photos of those plains, which they're already unofficially calling Sputnik Planum after the world's first man-made satellite.

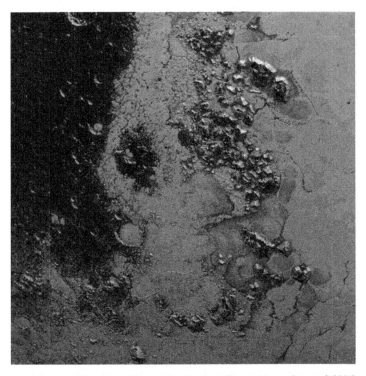

This image was taken by New Horizons' Long Range Reconnaissance Imager (LORRI) from a distance of 48,000 miles (77,000 kilometers). It shows a newly discovered mountain range near the southwestern margin of Pluto's Tombaugh Regio (Tombaugh Region), situated between bright, icy plains and dark, heavily-cratered terrain, July 24, 2015. (NASA/JHUAPL/SWRI via AP)

"Have a look at the icy frozen plains of Pluto," principal scientist Alan Stern said during a briefing at NASA headquarters. "Who would have expected this kind of complexity?"

Stern described the pictures coming down from 3 billion miles away as "beautiful eye candy."

"I'm still having to remind myself to take deep breaths," added Jeff Moore, head of the New Horizons geology team at NASA's Ames Research Center in California. "I mean, the landscape is just astoundingly amazing."

Spanning hundreds of miles, the plains are located in the prominent, bright, heart-shaped area of Pluto. Like the mountains unveiled Wednesday, the plains look to be a relatively young 100 million years old - at the most. Scientists speculate internal heating - perhaps from icy volcanoes or geysers- might still be shaping these crater-free regions.

"This could be only a week old for all we know," Moore said. He stressed that scientists have no hard evidence of erupting, geyser-like plumes on Pluto - yet.

Another possibility could be that the terrain, like frozen mud cracks on Earth, formed as a result of contraction of the surface.

The plains - which include clusters of smooth hills and fields of small pits - are covered with irregular-shaped, or polygon, sections that look to be separated by troughs. Each section is roughly 12 miles across.

The height of the hills is not yet known, nor their origin. It could be the hills were pushed up from below, or are knobs surrounded by eroded terrain, according to Moore. The fields of pits resemble glacial fields on Earth.

As of Friday's news conference, New Horizons was just over 2 million miles past Pluto and operating well. The spacecraft on Tuesday became the first visitor to the 4.5 billion-year-old Pluto, sweeping within 7,700 miles of its icy surface after a journey of 9½ years. It represented the last planetary stop on NASA's grand tour of the solar system, begun a half-century ago.

"I'm a little biased, but I think the solar system saved the best for last," Stern, a Southwest Research Institute planetary scientist, told reporters.

On Wednesday - just one day after the historic flyby - Stern and his team unveiled zoom-in photos showing 11,000-foot mountain ranges on Pluto, akin to the Rockies here on Earth. The plains are the mountains' neighbors to the north. The peaks are now known, informally at least, as the Norgay Montes. Tenzing Norgay was the

Sherpa guide for Sir Edmund Hillary when they conquered Mount Everest in 1953.

The huge, encompassing heart-shaped region already bears the last name of Clyde Tombaugh, the late American astronomer who discovered Pluto in 1930.

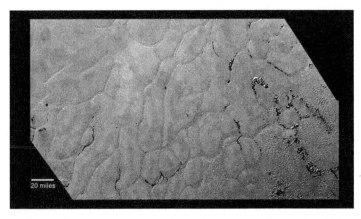

Close-up image from the heart-shaped feature on the surface of Pluto that reveals a vast, craterless plain, July 14, 2015. (NASA/JHUAPL/SWRI via AP)

New Horizons' science team promised Friday that the data will allow them to produce elevation maps of both Pluto and its big moon Charon.

It will take 16 months to transmit to Earth all the data collected during the close encounter. The Johns Hopkins University Applied Physics Laboratory in Laurel, Maryland, is managing the $720 million mission, which began with a launch from Cape Canaveral, Florida, in 2006 - months before Pluto was demoted from a full-fledged planet.

"This is just a taste of what I'm sure is in the unsent data" yet to come, Moore said.

'A DREAM COME TRUE'
Cape Canaveral, Florida, Saturday, July 25, 2015

Pluto is hazier than scientists expected and appears to be covered with flowing ice.

The team responsible for the New Horizons flyby of Pluto last week released new pictures Friday of the previously unexplored world on the edge of the solar system.

"If you're seeing a cardiologist, you may want to leave the room," principal scientist Alan Stern teased at the opening of the news conference at NASA headquarters. "There are some pretty mind-blowing discoveries."

This photo shows the atmosphere and surface features of Pluto, lit from behind by the sun. It was made 15 minutes after the New Horizons' spacecraft's closest approach, July 14, 2015. (NASA/JHUAPL/SwRI via AP)

NASA's New Horizons spacecraft, now 7.5 million miles beyond Pluto, has detected layers of haze stretching 100 miles (160 kilometers) into the atmosphere, much higher than anticipated. All this haze is believed to account for the dwarf planet's reddish color.

If you were standing on Pluto and looking up, you probably wouldn't notice the haze, said George Mason University's Michael Summers. In fact, New Horizons had to wait until after its closest approach on July 14, so the sun would silhouette Pluto and the atmosphere could be measured by means of the scattered sunlight.

As for the ice flows, they appear to be relatively recent: no more than a few tens of millions of years, according to William McKinnon of Washington University in St. Louis. That compares with the 4.5 billion-year age of Pluto and the rest of the solar system.

To see evidence of such recent activity, he said, is "simply a dream come true."

Temperatures on Pluto are minus 380 degrees Fahrenheit (minus 229 degrees Celsius), and so water ice would not move anywhere in such extreme cold. But McKinnon said the nitrogen and other ices believed to be on Pluto would be geologically soft and therefore able to flow like glaciers on Earth.

Some of that plutonian ice seems to have emptied into impact craters, creating ponds of frozen nitrogen. One of those semi-filled craters is about the size of metropolitan Washington D.C., McKinnon said.

These latest findings support the theory that an underground ocean might exist deep beneath Pluto's icy crust, McKinnon said.

These ice flows - which might still be active - are found on Pluto's vast icy plain, now called Sputnik Planum after Earth's first man-made satellite. The plain is about the size of Texas and occupies the left side of Pluto's bright heart-shaped feature, named Tombaugh Regio after the late astronomer who discovered Pluto in 1930, Clyde Tombaugh.

Sardar Tenzing Norgay, right, of Nepal and Edmund P. Hillary of New Zealand, left, show the kit they wore when conquering Mount Everest, on May 29, 1953 at the British Embassy in Katmandu, Nepal, June 26, 1953. (AP Photo)

It's evident now that the two "lobes" of the heart are quite different; Stern speculated that nitrogen snow could possibly be blowing from the brighter left, or western, side to the right.

One of Pluto's newly discovered mountain ranges now bears the name of Sir Edmund Hillary, who along with Sherpa guide Tenzing Norgay conquered Mount Everest in 1953. The New Horizons team already had named another series of mountains after Norgay.

The spacecraft traveled 3 billion miles over 9½ years to get the first close-up look of Pluto. The New Horizons team stressed that most of the collected data are still aboard the spacecraft and will take more than a year to obtain. Over the next several weeks, much of the incoming transmissions will consist of engineering or other technical data - and only a few images.

But starting in mid-September, "the spigot opens again," promised Stern, a scientist at the Southwest Research Institute. From then until fall 2016, "The sky will be raining presents with data from the Pluto system. It's going to be quite a ride."

A DAY ON PLUTO AND CHARON
Cape Canaveral, Florida, Saturday, November 20, 2015

NASA's newest Pluto pictures depict an entire day on the dwarf planet.

The space agency released a series of 10 close-ups of the frosty, faraway world Friday, representing one full rotation, or Pluto day. A Pluto day is equivalent to 6.4 Earth days.

The New Horizons spacecraft snapped the pictures as it zoomed past Pluto in an unprecedented flyby in July. Pluto was between 400,000 and 5 million miles from the camera for these photos.

A similar series of shots were taken of Pluto's jumbo moon, Charon. But the Pluto pictures stand out much more because of the orb's distinct heart-shaped region. Scientists call the heart Tombaugh Regio, after the U.S. astronomer who discovered Pluto in 1930.

New Horizons is now headed to a new target.

A series of 10 New Horizon close-ups of Pluto, representing one full rotation, or Pluto day. A Pluto day is equivalent to 6.4 Earth days, November 20, 2015. (NASA via AP)

New Horizon images of Charon representing one full rotation. Charon – like Pluto – rotates once every 6.4 Earth days, November 20, 2015. (NASA via AP)

Chapter 7

PLUTO'S MOONS

Illustration depicting Pluto and its five moons from a perspective looking away from the sun. It is adapted from a classic Voyager I montage of Jupiter's Galilean moons, and is intended to highlight similarities between the Pluto and Jupiter systems when adjusted for size. Approaching the system, the outermost moon is Hydra, seen in the bottom left corner. The other moons are roughly scaled to the sizes they would appear from this perspective, although they are all enlarged relative to the planet. (NASA/JPL/Mark Showalter, SETI Institute via AP)

NIX AND HYDRA
Los Angeles, Wednesday, June 21, 2006

Meet the newest kids in the solar system: Nix and Hydra.

The pair of moons orbiting Pluto were officially christened last week by the International Astronomical Union, which is in charge of approving celestial names.

Until last year, scientists thought Pluto was accompanied by only one moon, Charon. But the Hubble Space Telescope spotted the two satellites more than twice as far away as Charon and many times fainter.

The duo had been known by the tongue-twisting names S/2005 P 2 and S/2005 P 1. Earlier this year, the moons' discoverers, led by Alan Stern of the Southwest Research Institute in Boulder, Colorado, submitted their choices to the IAU.

The names, with roots in Greek mythology, were selected in part because their first letters, "N" and "H," were a tribute to the New Horizons spacecraft, Stern said Wednesday.

New Horizons blasted off earlier this year on a nine-year mission to study Pluto, the last unexplored planet in the solar system. Stern is the mission's principal investigator.

Nix was originally spelled "Nyx" by Stern's group. Nyx is the Greek goddess of darkness and Hydra is the nine-headed serpent that guarded the underworld. Pluto is the Roman god of the underworld.

But since a near-Earth object was already called Nyx, the IAU decided to tweak the spelling to "Nix" to avoid confusion.

Stern said he wasn't disappointed by the spelling change because the pronunciation and significance of the names were still intact.

"The joke was that they nixed Nyx," Stern said.

This summer, the IAU will debate whether Pluto should remain a planet. The discovery of an icy object slightly larger than Pluto in the Kuiper Belt last year reinvigorated the argument over whether to demote Pluto or add other planets.

NAMING PLUTO'S MINI MOONS
Cape Canaveral, Florida, Monday, February 11, 2013

Want to name Pluto's two tiniest moons? Then you'll need to dig deep into mythology.

Astronomers announced a contest Monday to name the two itty-bitty moons of Pluto discovered over the past two years.

Pluto is the Roman equivalent of the Greek's Hades, lord of the underworld, and its three bigger moons have related mythological names: Charon, the ferryman of Hades; Nix for the night goddess; and the multi-headed monster Hydra.

The two unnamed moons no more than 15 to 20 miles across need similarly shady references. Right now, they go by the bland titles of P4 and P5.

Among the choices: Hercules, the hero who slew Hydra; Obol, the coin put in the mouths of the dead as payment to Charon; Cerebrus, the three-headed dog guarding the gates of the underworld; Orpheus, the musician and poet who used his talents to get his wife, Eurydice, out of the underworld only to lose her by looking back: Eurydice; and Styx, the river to the underworld.

As of Monday afternoon, Styx and Cerebrus were leading. The vote tally is updated hourly.

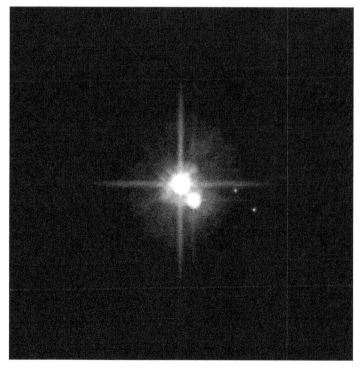

Pluto and three of its moons. A team of scientists using the Hubble Space Telescope said they discovered the tiniest moon yet around Pluto. That brings the number of known moons to five, February 22, 2006. (AP Photo/NASA)

"The Greeks were great storytellers, and they have given us a colorful cast of characters to work with," said Mark Showalter, senior research scientist at SETI Institute's Carl Sagan Center in Mountain View, Calif.

He and other astronomers who discovered the two mini-moons using the Hubble Space Telescope will make the winning selections.

Write-in name suggestions are welcomed, but they need to come from Greek or Roman mythology and deal with the underworld.

The name for the planetoid, or dwarf planet near the outer fringes of the solar system came from a little English girl. Pluto's discoverer, Clyde Tombaugh, liked that the first two letters were the same as the initials of late American astronomer Percival Lowell. Tombaugh discovered Pluto in 1930 using the Lowell Observatory in Flagstaff, Arizona.

NASA's New Horizons spacecraft is en route to Pluto, arriving in 2015 on the first robotic flyby ever of the planetoid.

The winning moon names will need final approval by the International Astronomical Union.

Hopefully, there won't be any conflicts like when the name Nix was picked. The night goddess actually is spelled Nyx, but an asteroid already had the moniker so the proper spelling for the moon had to be nixed.

STYX AND KERBEROS
Cape Canaveral, Florida, Monday, February 25, 2013

"Star Trek" fans, rejoice.

An online vote to name Pluto's two newest, itty-bitty moons is over. And No. 1 is Vulcan, a name suggested by actor William Shatner, who played Capt. Kirk in the original "Star Trek" TV series.

Vulcan snared nearly 200,000 votes among the more than 450,000 cast during the two-week contest, which ended Monday. In second place with nearly 100,000 votes was Cerberus, the three-headed dog that guarded the gates of the underworld.

Vulcan was the Roman god of lava and smoke, and the nephew of Pluto. Vulcan was also the home planet of the pointy-eared humanoids in the "Star Trek" shows. Think Mr. Spock.

"174,062 votes and Vulcan came out on top of the voting for the naming of Pluto's moons. Thank you to all who voted!" Shatner said in a tweet once the tally was complete.

Actor Leonard Nimoy, who portrayed the reason-and-logic-based Spock, had this to say in an email to The Associated Press: "If my people were emotional they would say they are pleased."

Leonard Nimoy poses for a photograph in Los Angeles, August 8, 2006. (AP/Ric Francis)

Don't assume Vulcan and Cerberus are shoo-ins, though, for the two tiny moons discovered over the past two years with the Hubble Space Telescope.

The contest was conducted by SETI Institute in Mountain View, Calif., the research base for the primary moon hunter. The 10 astronomers who made the discoveries will take the voting results into account, as they come up with what they consider to be the two best names.

The International Astronomical Union has the final say, and it could be another month or two before an edict is forthcoming. Now known as P4 and P5, the moons are 15 to 20 miles across.

The leader of the teams that discovered the mini-moons, Mark Showalter said Monday he is leaning toward the popular vote.

But Showalter pointed out that asteroids thought to orbit close to the sun are called vulcanoids, and there could be some confusion if a moon of Pluto were to be named Vulcan. Vulcan, in fact, was the name given in the 19th century to a possible planet believed to orbit even closer to the sun than Mercury; no such planet ever was found.

What's more, Showalter said in a phone interview, Vulcan is associated with lava and volcanoes, while distant Pluto is anything but hot.

As for Cerberus, an asteroid already bears that name, so maybe the Greek version, Kerberos, would suffice, said Showalter, a senior research scientist at SETI's Carl Sagan Center.

Styx landed in No. 3 position with nearly 88,000 votes. That's the river to the underworld.

Pluto's three bigger moons are Charon, Nix and Hydra.

To be considered, the potential names for the two mini-moons also had to come from Greek or Roman mythology, and deal with the underworld. Twenty-one choices were voted on. Of those, nine were write-in candidates suggested by the public, including Shatner's entry for Vulcan.

Shatner's second choice for a name, Romulus, did not make the cut. That's because an asteroid already has a moon by that name along with a moon named Remus.

And forget the Disney connection.

"We love Mickey, Minnie and Goofy, too," Showalter informed voters a few days into the voting. "However, these are not valid names for astronomical objects. Sorry."

Altogether, 30,000 write-in candidate names poured in.

Showalter said he will keep the list handy as more moons undoubtedly pop up around Pluto once NASA's New Horizons spacecraft arrives in 2015. It will be the first robotic flyby ever of the planetoid, or dwarf planet near the outer fringes of the solar system.

"I have learned not to underestimate Pluto," Showalter wrote on the website. With so many good names available, "Pluto needs more moons!"

COSMIC DANCE
Washington, Wednesday, June 3, 2015

There's a chaotic dance going on at the far end of our solar system, involving Pluto and five of its closest friends, a new study finds.

Hubble Space Telescope images of Pluto, its largest moon Charon, and tinier moons Styx, Nix, Hydra and Kerberos show the odd rhythmic gyrations of the six distant objects in a dance unlike anything in our solar system.

What makes it so odd is that there's a double set of dances going on. First, Pluto and Charon are locked together in their own waltz "as if they are a dumbbell" with a rod connecting them, said study author Mark Showalter of the SETI Institute in California. It's the solar system's only binary planet system, even though Charon isn't technically a planet, he said. Pluto, too, is no longer considered a full planet.

This combination image released by NASA shows Pluto, left, and its moon, Charon, with differences in surface material and features depicted in exaggerated colors made by using different filters on a camera aboard the New Horizons spacecraft. In this composite false-color image, the apparent distance between the two bodies has also been reduced, July 13, 2015. (NASA/APL/SwRI via AP)

"It's pretty darn weird," Showalter said.

But Pluto and Charon aren't alone, and that's where it gets more complicated.

The four little moons circle the Pluto-Charon combo, wobbling a bit when they go closer to either Pluto or Charon, being pushed and pulled by the two bigger objects.

Those four moons orbit Pluto-Charon in a precise rhythmic way, but with a twist: They also interact when they near each other. So it seems like they all dance to one overarching beat but not quite in the same way, just doing their own thing, said planetary scientist

Heidi Hammel of the Association of Universities for Research in Astronomy.

"It's kind of like you'd see at a Grateful Dead concert," Hammel said. She wasn't part of the study, but praised it as giving a glimpse of what might be happening in other distant star systems where there are two stars and planets that revolve around them, like the mythical Star Wars world of Tatooine.

With the tiny moons wobbling and flipping over in an unpredictable and chaotic way, if you lived on Nix or Hydra, the sun would come up in different parts of the sky, if at all on some days, Showalter said.

"It's a very strange world," he said. "You would literally not know if the sun is coming up tomorrow."

NASA's $700 million New Horizons spacecraft will arrive in the Pluto system in mid-July after a nine-year 3 billion mile flight that started before Pluto was demoted to dwarf-planet status.

SPECTACULAR CHARON
Cape Canaveral, Florida, Thursday, October 1, 2015

Pluto's big moon, Charon, is being revealed in all its rugged glory.

NASA released the best color pictures yet of Charon on Thursday. The images were taken by the New Horizons spacecraft during its flyby of Pluto in July and transmitted to Earth 1½ weeks ago.

Massive canyons and fractures are clearly visible on Charon, which is more than half of Pluto's size. The canyons stretch more than 1,000 miles (1,600 kilometers) above the equator, across the entire face of Charon. These canyons are four times as long as the Grand Canyon and, in places, twice as deep.

Even better pictures are anticipated as flight controllers at Johns Hopkins University receive more data from New Horizons, now about 3 billion miles (5 billion kilometers) from Earth.

Charon, in enhanced color captured by NASA's New Horizons spacecraft just before closest approach on July 14, 2015, October 1, 2015. (NASA/JHUAPL/SwRI via AP)

Chapter 8

MORE TO COME IN THE KUIPER BELT

Pluto-Kuiper Express spacecraft, artist conception (NASA/AP)

SURPRISINGLY DIVERSE LANDSCAPE
Cape Canaveral, Florida, Monday, September 11, 2015

The spigot has opened again, and Pluto pictures are pouring in once more from NASA's New Horizons spacecraft.

These newest snapshots reveal an even more diverse landscape than scientists imagined before New Horizons swept past Pluto in July, becoming the first spacecraft to ever visit the distant dwarf planet.

"If an artist had painted this Pluto before our flyby, I probably would have called it over the top - but that's what is actually there," said Alan Stern, New Horizons' principal scientist from Southwest Research Institute in Boulder, Colorado.

In one picture, dark ancient craters border much younger icy plains. Dark ridges also are visible that some scientists speculate might be dunes.

One outer solar-system geologist, William McKinnon of Washington University in St. Louis, said if the ridges are, in fact, dunes, that would be "completely wild" given Pluto's thin atmosphere.

"Either Pluto had a thicker atmosphere in the past, or some process we haven't figured out is at work. It's a head-scratcher," McKinnon said in a written statement.

The jumble of mountains, on the other hand, may be huge blocks of ice floating in a softer, vast deposit of frozen nitrogen.

This photo shows a synthetic perspective view of Pluto, based on the latest high-resolution images to be downlinked from NASA's New Horizons spacecraft. The new close-up images of Pluto reveal an even more diverse landscape than scientists imagined, July 14, 2015. (NASA/Johns Hopkins University Applied Physics Laboratory/Southwest Research Institute via AP)

After several weeks of collecting engineering data from New Horizons, scientists started getting fresh Pluto pictures last weekend. The latest images were released Thursday.

This photo shows a 220-mile (350-kilometer) wide view of Pluto, July 14, 2015. (NASA/Johns Hopkins University Applied Physics Laboratory/Southwest Research Institute via AP)

Besides geologic features, the images show that the atmospheric haze surrounding Pluto has multiple layers. What's more, the haze crates a twilight effect that enables New Horizons to study places on the night side that scientists never expected to see.

Monday marks two months from New Horizons' close encounter with Pluto on July 14, following a journey from Cape Canaveral, Florida, spanning 3 billion miles and 9½ years. As of Friday, the spacecraft was 44 million miles past Pluto.

So much data were collected during the Pluto flyby that it will take until next fall to retrieve it all here on Earth. The spacecraft is operated from the Johns Hopkins University Applied Physics Laboratory in Laurel, Maryland, which also designed and built it.

New Horizons' next target, pending formal approval by NASA, will be a much smaller object that orbits 1 billion miles beyond Pluto. It, too, lies in the so-called Kuiper Belt, a frigid twilight zone on the outskirts of our solar system. Following a set of maneuvers,

New Horizons would reach PT1 - short for Potential Target 1 - in 2019.

'MAKES YOU FEEL LIKE YOU ARE THERE'
Cape Canaveral, Florida, Thursday, September 17, 2015

The newest pictures of Pluto are so up-close and personal that the mission's top scientist says it "makes you feel you are there."

NASA released the photos Thursday. The images were gathered by the New Horizons spacecraft that swept past the dwarf planet in July.

Pluto's curvature is featured in the latest pictures, with the sun providing dramatic backlighting. Rugged terrain is shown, as is the extended atmosphere of the tiny orb on the frigid outskirts of the solar system. The panorama stretches 780 miles.

This photo shows the atmosphere and surface features of Pluto, lit from behind by the sun. It was made 15 minutes after the New Horizons' spacecraft's closest approach, July 14, 2015. (NASA/JHUAPL/SwRI via AP)

Principal scientist Alan Stern says the pictures shed new light on Pluto's mountains, glaciers and plains.

Johns Hopkins University operates New Horizons, the world's first visitor to Pluto. The spacecraft is now 48 million miles beyond Pluto.

BLUE SKY AND RED ICE
Cape Canaveral, Florida, Thursday, October 8, 2015

The sky over Pluto may not be sunny but it's undoubtedly blue.

NASA's New Horizons spacecraft discovered Pluto's blue sky during the historic flyby of the icy dwarf planet in July. The images of Pluto's atmospheric haze were beamed down last week and released by NASA on Thursday.

This New Horizons Ralph/Multispectral Visible Imaging Camera (MVIC) image shows the blue color of Pluto's haze layer. The high-altitude haze is thought to be similar in nature to that seen at Saturn's moon Titan. This image was generated by software that combines information from blue, red and near-infrared images to replicate the color a human eye would perceive as closely as possible, October 8, 2015. (NASA/JHUAPL/SwRI via AP

The particles in the atmospheric haze are actually red and gray, according to scientists. But the way the particles scatter blue light is what has everyone excited about the dwarf planet orbiting on the far fringes of our solar system, a twilight zone known more formally as the Kuiper Belt.

"Who would have expected a blue sky in the Kuiper Belt? It's gorgeous," Alan Stern, the principal scientist for New Horizons, said in a NASA release about the latest images.

The blue tint can help scientists understand the size and makeup of the haze particles surrounding Pluto, where twilight constantly reigns given the 3.6 billion-mile distance between it and the sun.

Pluto's high-altitude haze seems to be comparable to that of Saturn's moon, Titan, according to NASA, and the result of interaction between molecules.

In another finding Thursday, scientists have uncovered numerous ice patches on Pluto's surface. The exposed water ice appears to be, mysteriously, red.

Scientists said they are uncertain why the ice appears in certain places at Pluto and not others.

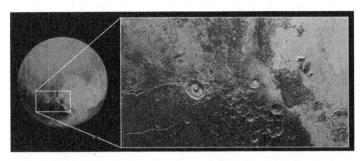

This image released by NASA shows regions with exposed water ice highlighted in blue in this composite image taken with the New Horizons spacecraft's Ralph instrument. The image combines visible imagery from the Multispectral Visible Imaging Camera (MVIC) with infrared spectroscopy from the Linear Etalon Imaging Spectral Array (LEISA). The scene is approximately 280 miles (450 kilometers) across, October 8, 2015. (NASA/JHUAPL/SwRI via AP)

Launched in 2006, New Horizons is now 63 million miles beyond Pluto. Johns Hopkins University in Maryland is operating the spacecraft for NASA.

CLOSE ENCOUNTER WITH MU69 IN 2019
Cape Canaveral, Florida, Thursday, October 22, 2015

The spacecraft that gave us the first close-up views of Pluto now has a much smaller object in its sights.

NASA's New Horizons was programmed to fire its thrusters Thursday afternoon, putting it on track to fly past a recently discovered, less than 30-mile-wide object out on the solar system frontier. The close encounter with what's known as 2014 MU69 would occur in 2019. It orbits nearly 1 billion miles (1.6 billion kilometers) beyond Pluto.

Flight controllers at the Johns Hopkins University Applied Physics Laboratory in Laurel, Maryland, sent commands in advance for the course change. In all, four maneuvers will be needed, carried out over the next two weeks. Thursday afternoon's was the first; it was expected to be several hours before controllers received confirmation that everything had gone well.

Launched in 2006, New Horizons became Pluto's first visitor from planet Earth in July. The spacecraft remains in excellent health following a 3 billion-mile (4.8 billion kilometer) journey and still holds a year's worth of scientific data for transmission back to Earth.

This image shows a combination of images captured by the New Horizons spacecraft with enhanced colors to show differences in the composition and texture of Pluto's surface. The images were taken when the spacecraft was 280,000 miles (450,000 kilometers) away. The New Horizons was programmed to fire its thrusters Thursday, October 22, 2015, putting it on track to fly past a recently discovered, less than 30-mile-wide object out on the solar system frontier. The close encounter with 2014 MU69 would occur in 2019. It orbits nearly 1 billion miles beyond Pluto, July 24, 2015. (NASA/JHUAPL/SwRI via AP)

NASA and the New Horizons team chose 2014 MU69 in August as New Horizons' next potential target, thus the nickname PT-1.

Like Pluto, MU69 orbits the sun in the frozen, twilight zone known as the Kuiper Belt.

The extremely remote, faint object was spotted by the Hubble Space Telescope in 2014. It beat out a few other candidates because it will take less fuel to get there.

MU69 is thought to be 10 times larger and 1,000 times more massive than average comets, including the one being orbited right now by Europe's Rosetta spacecraft. On the other end, MU69 is barely 1 percent the size of Pluto and perhaps one-ten-thousandth the mass of the dwarf planet. So the new target is a good middle ground, according to scientists.

Lead scientist Alan Stern said flight controllers still are working out just how close New Horizons will be able to zoom past MU69. The goal is to get closer than the 7,770 miles (12,500 billion kilometers) that the spacecraft came within Pluto.

The team plans to formally ask NASA next year to fund the mission extension for studying MU69. Scientists promise a better name before showtime on January 1, 2019.

"Although this flyby probably won't be as dramatic as the exploration of Pluto we just completed," Stern wrote in a blog earlier this month for Sky & Telescope magazine, "it will be a record-setter for the most distant exploration of an object ever made."

Johns Hopkins designed the spacecraft, about the size of a baby grand piano, and has been operating it for NASA.

EPILOGUE

MISSION TO PLUTO
Associated Press Interactive, Friday, July 10, 2015

New Horizons Spacecraft

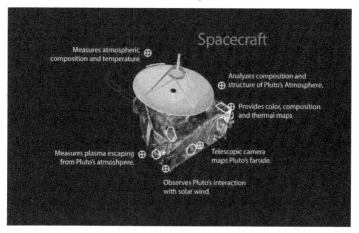

Ten Year Voyage to Pluto

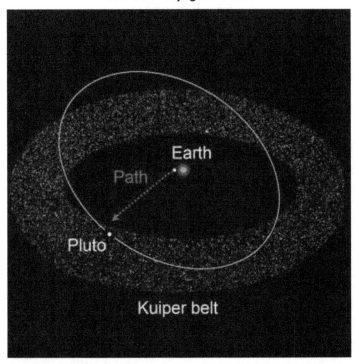

Launch: The first 13 months included spacecraft and instrument checkouts, instrument calibrations, small trajectory correction maneuvers and rehearsals for the Jupiter encounter. New Horizons passed the orbit of Mars on April 7, 2006. It also tracked a small asteroid, later named "APL", in June 2006.

Jupiter Encounter: Closest approach occurred February 28, 2007. Moving about 51,000 miles per hour, New Horizons flew about 3 to 4 times closer to Jupiter than the Cassini spacecraft, coming within 1.4 million miles of the large planet.

During the journey, New Horizons also crossed the orbits of Saturn (June 8, 2008), Uranus (March 18, 2011), and Neptune (August 25, 2014).

AP Editions

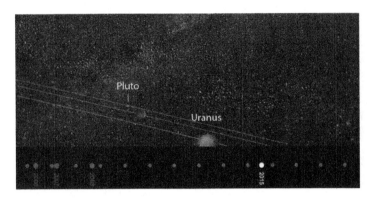

New Horizons observations of the Pluto system were planned well in advance, with each instrument programmed to perform specific functions as the spacecraft speeds past. Pluto is about 40 times farther out from the Sun than Earth.

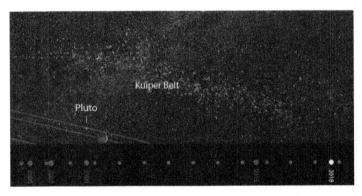

The spacecraft is programmed to fly beyond Pluto and explore the Kuiper Belt with a communications system and scientific instruments that can work in much lower sunlight.

New Horizons Spacecraft Reaches the Outer Planets

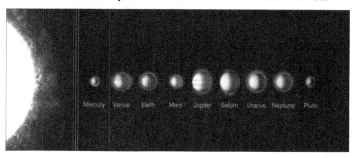

Planets in our solar system from closest to furthest from the Sun: Mercury, Venus, Earth, Mars, Jupiter, Saturn, Uranus, Neptune, and Pluto

Mercury

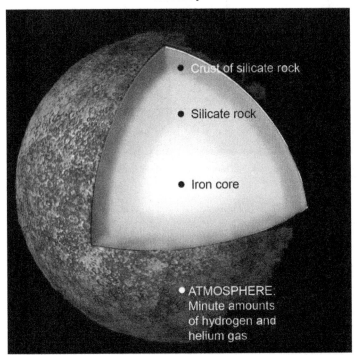

- Crust of silicate rock
- Silicate rock
- Iron core
- ATMOSPHERE: Minute amounts of hydrogen and helium gas

Craters on Mercury range from small bowl-shaped cavities to multi-ringed impact basins hundreds of kilometers across.

Venus

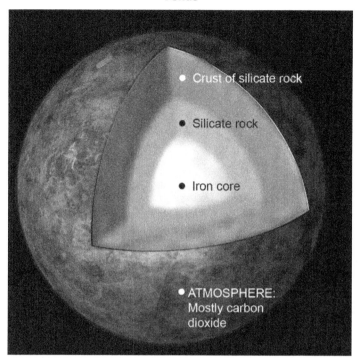

- Crust of silicate rock
- Silicate rock
- Iron core
- ATMOSPHERE: Mostly carbon dioxide

Venus with the Sun behind. Venus is the hottest planet, though not the closest to the Sun.

Earth

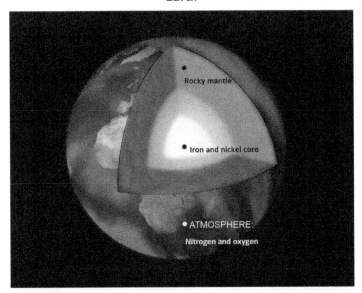

Clouds can be seen high above Earth's atmosphere.

Mars

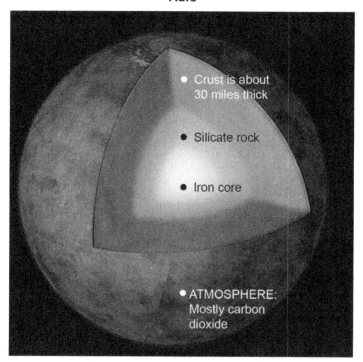

- Crust is about 30 miles thick
- Silicate rock
- Iron core
- ATMOSPHERE: Mostly carbon dioxide

Surface shows both wind and water erosion.

Jupiter

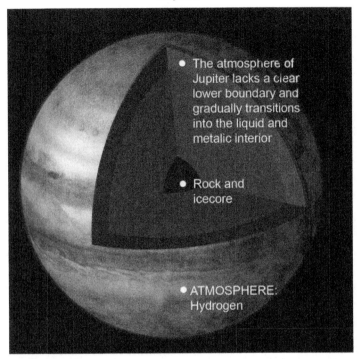

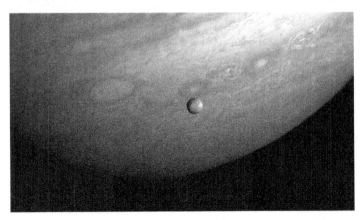

The moon Io orbiting Jupiter. Io is the most volcanically active body in the solar system.

Saturn

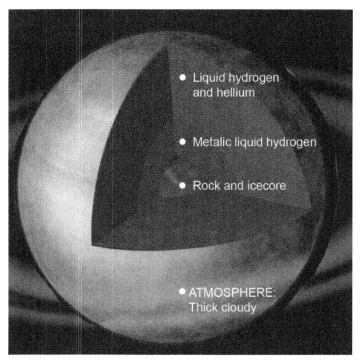

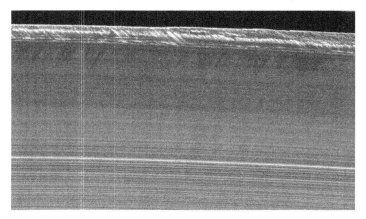

Ring particles are made almost entirely of water ice, with some rock.

Uranus

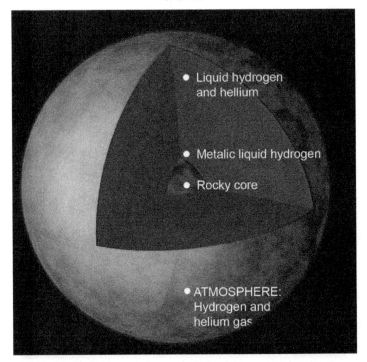

- Liquid hydrogen and hellium
- Metalic liquid hydrogen
- Rocky core
- ATMOSPHERE: Hydrogen and helium gas

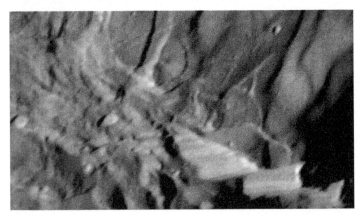

Miranda, one of 27 known moons of Uranus, has a very sculpted surface.

Neptune

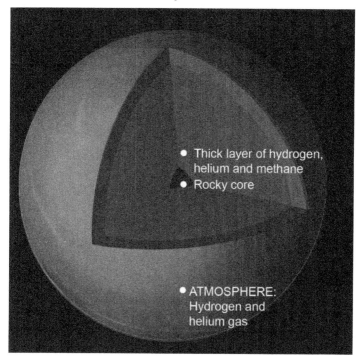

- Thick layer of hydrogen, helium and methane
- Rocky core

- ATMOSPHERE: Hydrogen and helium gas

Cirrus-type clouds floating high in Neptune's atmosphere.

Pluto

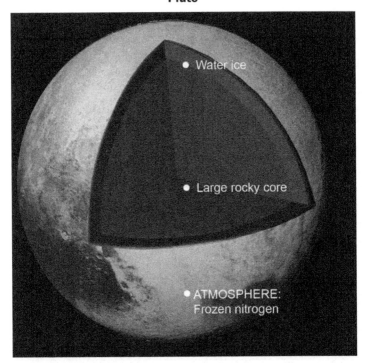

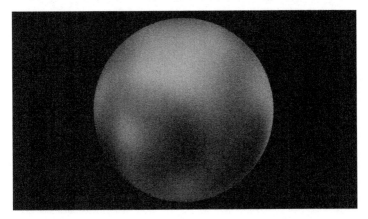

Composite showing the most detailed images of Pluto prior to the New Horizons flyby.

CITATIONS AND BYLINES

THE AP EMERGENCY RELIEF FUND

When Hurricane Katrina hit the Gulf Coast in 2005, many Associated Press staffers and their families were personally affected. AP employees rallied to help these colleagues by setting up the AP Emergency Relief Fund, which has become a source of crucial assistance for the past 10 years.

Established as an independent 501(c)(3), the Fund helps AP staffers who have suffered damage or losses as a result of conflict or natural disasters. These grants are used to rebuild homes, move to safe houses and repair and replace bomb-damaged belongings.

The AP matches all gifts in full and also donates the net proceeds from AP Essentials, AP's company store, to the Fund.

HOW TO GIVE

In order to be ready to help the moment emergencies strike, the Fund relies on the generous and ongoing support of the extended AP community. All donations are matched in full by The Associated Press and can be made any time at http://www.ap.org/relieffund and are tax deductible.

On behalf of the AP staffers and families who receive aid in times of crisis, the AP Emergency Relief Fund Directors and Officers thank you.

ALSO AVAILABLE FROM AP EDITIONS

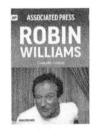

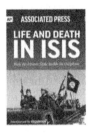

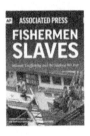

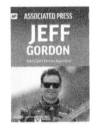

CPSIA information can be obtained at www.ICGtesting.com
Printed in the USA
BVOW06s2118180216

437270BV00003B/9/P